LEARN MATHEMATICS SYMBOLS, DEFINITIONS, AND LANGUAGE

(Volume 1 of a Math Learning Series)

By Ardin Marschel

I hope you find this book useful to students learning Math Basics.

Any Comments, Suggestions, and Yes, especially Criticisms; will be gratefully appreciated by the author.

Author's E-mail address: Ardin2000@aol.com

StudyKat Publishing, Myrtle Beach, SC

Printed and Bound in United States of America

First Edition

Contact information: Ardin2000@aol.com

ISBN: 978-0-615-86015-2

Table of Contents

Dedications:

To my intelligent and beautiful wife, Susan Hanley, who also gave great help to review and gave editing suggestions for all parts of this book.

Added Acknowledgements:

Help with reviewing the book contents: Errick Taylor, James Anderson, and Jessica Gonsalves

Help with designing the cover: Kaldin Taylor, Errick Taylor, Tori Taylor, Frank Taylor, Jr., Julia Taylor, Gavin Taylor, Steve Gonsalves, Julia Gonsalves, Jessica Gonsalves, Matt Gonsalves, Frank Taylor, Sr., Mrs. Frank Taylor, Sr., Bob Dempsey, and Kathy Dempsey

Help with designing the StudyKat logo: Jessica Gonsalves and Julia Gonsalves

PREFACE:

WHO WOULD BENEFIT FROM THIS BOOK:

1) High School students wishing to improve grades in Mathematics classes
2) High School students preparing for standard tests for advancing to the next higher grade
3) High School students brushing up on Mathematics preparing for SAT or ACT tests
4) Parents of High School students who want to help their kids with Mathematics
5) Adults who want to complete a GED (General Education Diploma)
6) College Students brushing up on Mathematics before College courses
7) Students of all ages who wish to have a quick reference guide for Math symbols and terms

This book serves as a tool or basic book for understanding the language and symbols used in Mathematics and Arithmetic for solving problems, especially fearsome Math Word Problems.

To begin, Arithmetic is the study of the simplest ways of working with numbers; such as: adding, subtracting, multiplying, and dividing.

Within a School education system, after mastering basic Arithmetic, students mostly move on to at least some of the many other topics of Mathematics. These added and more complicated topics are usually needed to understand most of the Sciences for College and many jobs. Also, many of the basic topics in Arithmetic and Mathematics must be understood to get good grades in school on tests both in Mathematics as well as in Science courses.

Some adults are not able to complete High School either in the USA or a similar school in another country. Later on in life these adults may find that they need the High School type of education. One way to get this education for these adults is to study to pass a GED test to get a GED diploma or certificate. A GED diploma or certificate is understood by most employers as a good substitute for a High School Diploma. GED stands for General Education Development or General Education Diploma. A GED test contains five parts which are: 1) Language Arts of Writing, 2) Social Studies, 3) Science, 4) Language Arts of Reading, and 5) Mathematics. This book may help to understand the Mathematics to get a good grade for the part 5) Mathematics section of a GED test.

A large number of students in High School, as well as adults, want to also go to a college or university for added certificates or degrees. Generally colleges and universities require either a High School Diploma or a GED passing certificate. Also, entrance applications for colleges or universities in the USA most often want the grades or scores from either an SAT (Scholastic Assessment Test, also called the Scholastic Aptitude Test) or ACT (American College Test). Books like this one may help you to do better on all of these tests.

Learning Science, Arithmetic, and Mathematics needs to be done in steps starting from the beginning basics and then on to more complicated ideas. These steps of knowledge are usually taught in courses during each school year as well as adult education classes. Most of those school courses at higher

Grades or years assume that students know what ideas were given, or hopefully learned, in earlier Grades or years. In order to take more advanced topics or courses you should know what are called prerequisites. Prerequisites are ideas or knowledge that a student should already know before starting a class. What happens if a student starts a class and is weak on what should have already have been learned before? Such a student may fall behind and have a hard time keeping up with the teacher or other students in a class. Low grades may result.

Help for these students is needed. Here are some ways of getting help:

1) The teacher may give extra help, if the teacher has enough time.
2) Sometimes a student has a parent or a relative that can help.
3) Maybe getting free tutoring help is possible.
4) The student could decide to pay money for a tutor.
5) The student can study between classes on the books or other materials from the teacher.

OR

6) Books like this one may help by giving added information and explanations.

The plan for this book and future books in this series of Math Help books is to give extra help to understand basic Mathematics. These basics are also used in Science books and classes. Lots of information is in this book including introductory basics for the following and more:

Arithmetic
Number handling in Mathematics
Algebra
Geometry
Trigonometry
Probability and Statistics
Data graphing
Handling Scientific numbers

This book's content gives some important background knowledge to help a student to do well in Mathematics and Sciences courses. THIS book is the first in a series for learning Mathematics leading to more complicated topics in future books. The content below covers the language and symbols used in basic Arithmetic and Mathematics and introduces how they are usually communicated or written for efficiency as a type of shorthand. Understanding this language and symbols is needed for solving Arithmetic and Mathematics problems. This book contains basic help to increase your test scores in school, do better on tests, and maybe enjoy problem solving like the fun of solving a puzzle.

To emphasize certain very important ideas, "StudyKat" is placed nearby. Here is what "StudyKat" looks like as your Study helper Kat:

StudyKat

Happiness with Mathematics makes StudyKat "Purrrrrrrr".

PART I: Whole Number Symbols used in Arithmetic and Mathematics

A. Basic Introduction

The Arabic number symbol system is almost universally used in around the World, but is not the only number system in use. The Arabic number symbol system was originally called Hindu-Arabic, but mostly now is called simply the Arabic number system. The Hindu-Arabic number system actually originated in India. It uses base 10 numerals. A future book in this series of books explains what is meant by base 10 as well as other number systems like Binary, the base 2 system, used by computers. Numbers representing the same quantities of items look a little different in different number systems.

In Mathematics, symbols form a short-hand, or abbreviated, language between people. This book is an Introduction to the language or code of Arithmetic and Mathematics so as to be able to more easily read and understand it.

B. Symbols for Numbers

Arabic numbers are:

0, 1, 2, 3, 4, 5, 6, 7, 8, 9, and combinations of these numbers

Numbers also can be shown by other symbols, such as, Greek letters, for example: α, β, γ, δ, ε... as well as a few other common number symbol systems as shown in the following table. The table below lists some common number systems but these are not the only ones used in various parts of the World.

Here are listings of number symbols in various languages for the first 10 numbers:

Arabic Numbers: 1, 2, 3, 4, 5, 6, 7, 8, 9, 10...

Roman Numerals: I, II, III, IV, V, VI, VII, VIII, IX, X...

Binary Numbers: 1, 10, 11, 100, 101, 110, 111, 1000, 1001, 1010...

Greek Numerals (lower case): α, β, γ, δ, ε, ζ, η, θ, ι, κ...

Greek Numerals (upper case): Α, Β, Γ, Δ, Ε, Ζ, Η, Θ, Ι, Κ...

The Greek number symbols are also used in the Greek alphabet but are frequently used for either constants or variables. For example, the Greek symbol "θ" is commonly used to symbolize an angle between two intersecting straight lines in Geometry or Trigonometry. In Chemistry, the Greek symbols "α", "β", and "γ" are used to label parts in a chemical molecule. Roman numbers, sometimes called Roman numerals, are commonly used in lists to separate different sections of a book or number Super Bowl games.

But symbols like "θ", "α", "β", "γ" are only symbols and are used when their actual value is yet unknown or can be many different possible values. The question usually comes up: What is their actual number or value in a particular situation? Such actual values are commonly found when solving a Mathematics problem. Until an exact number value is "solved for" these symbols are called variables. A basic introduction to variables is in Appendix A after the Glossary in this Book.

PART II: Mathematical Symbols are Abbreviations used in Math Expressions or Statements.

Symbols, such as listed below, replace words or phrases and make expressions which are mathematical sentences. These expressions (or sentences) are usually shorter to read and communicate than either spoken expressions or lengthy wordings. Therefore, a math expression is just a sentence containing symbols that are shorthand for the words. There are many such words or Math operation symbols. Here is a list of the more common ones that are used.

Symbol - - Description in Words

+ - - Plus or Add

– - -Minus or Subtract

± - - Plus or Minus

× or • - - Multiply – Also, in many Math expressions, multiplication is indicated by placing a Number or Variable right next to another Number or Variable - For example: (3)(2) indicates to multiple (3) by (2); or; another example is that 3x indicates that the Number "3" times "x" is meant. That is, 3x equals 3 multiplied by x.

÷ or / or — - - Divide - Note that divide sign, "÷"; the slanted line "/"; or the underline "—"(in a fraction under the numerator and above the denominator); commonly separates numbers or symbols where the first or top number or item is divided by the second or bottom number or item. For example, 3/4 indicates 3 divided by 4; or; indicates a number called a fraction. See Appendix C after the Glossary near the end of this book for a basic introduction to Fractions.

= - - Equal or Equals

- - Abbreviation usually meaning "number"; or; a weight in pounds

% - - Percent or Percentage

< - - Less than – That is, A< B indicates that A is less than B.

<< - - Much less than or a lot less than – That is, A<< B indicates that A is much less or a lot less than B

> - - Greater than or more than - That is, A > B indicates that A is greater than B

>> - - Much greater or a lot greater or more than – That is, A >> B indicates that A is much or a lot greater or much more than B.

≤ - - Less than or equal to – That is, A ≤ B indicates that A is either less than B, or, A is equal to B; or; A is neither greater nor more than B. Sometimes the underline (that is, meaning equal) in this symbol is a full equal sign: = under the < symbol.

≥ - - Greater than or more than, or, equal to - That is, A ≥ B indicates that A is either greater than or more than B, or A is equal to B; or; A is not less than B. Sometimes the underline (that is, meaning equal) in this symbol is as full equal sign: = under the > symbol.

≠ - - Not equal – That is, A ≠ B indicates that A can be less than B, or, A can be greater or more than B, but A is not equal to B.

StudyKat

"Meow" A diagonal line through a symbol is used a lot or frequently, even on signs that we see on roads and other public places to indicate NOT whatever the underlying symbol is.

NOTE that a diagonal line through any symbol indicates the opposite of whatever the symbol means, or, deletion of the meaning of the symbol. See the above example, of "Not equal".

This NOTE decreases the need to include in this listing a duplication of the long list of symbols just to indicate the opposite of them with a diagonal line through them.

. - - Decimal point – used to separate the integer part from the decimal fraction part of a number – For example, 34.87 contains the integer "34" and the decimal fraction part ".87".

≈ - - Approximately equal

~ - - Approximately or Approximate or Similar

! - - Factorial (defined in the below Definitions Glossary)

**** or "superscripted number" or ↑** - - indicates raising a number to a power (exponent) – The power or exponent is written after this symbol. – For example: 2^3 equals 2 x 2 x 2. The power or exponent indicates how many times a number or item is repeated in a multiplication series.

: - - Ratio symbol usually placed between two numbers being compared - For example: 3:2 indicates that "3" is compared to "2".

| | - - Absolute value of the number between the upright parallel lines indicates: Positive numbers between the | | stay positive and negative numbers between the | | are changed to from being negative to positive; or; said another way: A numerical value or expression that is negative in number value within the symbol | | is multiplied by "-1" and therefore made positive but of the same number size. Yet another meaning for absolute value is the number distance of that which is between the upright lines | | from the value of zero. For example, both +4 and -4 are "4" units in number distance from zero. Therefore; both | +4 | equals +4, and, | - 4 | equals +4.

Δ - - Difference or Change of whatever is shown after the symbol - That is, Δx is the change in the value of x.

1/ - - Inverse of the number under the slanted line, also called reciprocal: For example, the reciprocal or inverse of 2 is ½.

∑ - - Summation or adding of a series of items that follow the ∑ symbol - For example: ∑ 2, 3, 4 equals 2 + 3 + 4 = 9

∏ - - Multiplication of a series of items that follow the ∏ symbol - For example: ∏ 2, 4 equals 2 x 4 = 8.

Lim - - Limit of whatever follows the symbol - Usually the Limit is determined to be the value that a variable quantity approaches closer and closer to as a defined endpoint value or other condition is very closely approached. For example: $\text{Lim}_{\Delta x \to 0} (2 + \Delta x) = 2$ - Another example: $\text{Lim}_{\Delta x \to 0} \Delta x = 0$

( ) - - Parentheses - These are symbols frequently used to isolate or group together a portion of a mathematical expression to only whatever is within the parentheses.

[ ] - - Brackets - These are symbols frequently used to isolate or group together a portion of a mathematical expression to only whatever is within the brackets.

{ } - - Braces - These are symbols frequently used to isolate or group together a portion of a mathematical expression to only whatever is within the braces.

x, y, z, t, ... - - Variables are symbols for quantities that are unknown or vary depending on problem conditions.

f(x, y, . . .) - - Function of x, y, or other numbers . . . ; or a function whose value depends on the values of the numbers or variables within the parentheses ( ) - For example: f(x) = 2x + 4 is a function f(x) that has a value determined by what "x" is. If x = 2, then f(x) = f(2) = 2 • 2 + 4 = 8. If x = 5, then f(x) = f(5) = 2 • 5 + 4 = 14. ** Note the more detailed definition of a function in the Glossary below.

Δx - - Change of the value of a variable, here symbolized by x. If x varied from 1 to 25 by integer values, then Δx may have values that may be anyone of 1, 2, 3 . . . up to 25. Sometimes the change of the variable includes fractional values such as 1.1, 1.2, 5.5, and so forth. For a specific example, if x changed from 1 to 5 in a particular situation, then Δx over the entire change in value would equal 4 which is how much x changed when it varies from 1 to 5. Sometimes, Δx may be called the span of x.

Infinity Symbol - -

$$\infty$$

|| - - Parallel

→ - - An arrow over a variable or number most often indicates a vector quantity, that is, a quantity with both a magnitude value and a direction. The value is called a magnitude or size or amount of the vector quantity. See also the definition of a vector definition in the below Glossary.

√ - - Root, usually means the square root if there is no number above the symbol (The square root of 4 is symbolized as √4). The √4 equals 2. The square root of a number is a smaller number which, when multiplied by itself, results in the number which is to the right of the √ symbol. In the above example, the square root of 4 is 2 because 2 multiplied by 2 equals 4. If there is a number above the symbol then that number is the power of the root. For example, a cube root would have a "3" above the root symbol. The cube root of 27 is symbolized as ^{3}√27 and equals 3. That is, 3 x 3 x 3 equals 27.

Root Symbol Examples:

√ or $\sqrt{\ }$

√4 or $\sqrt{4}$ or ^{2}√4 or $\sqrt[2]{4}$

^{3}√27 or $\sqrt[3]{27}$

The following Trigonometry terms apply to right (90 degree or 90^0) triangles:

A right (90 degree or 90^0) triangle is shown here:

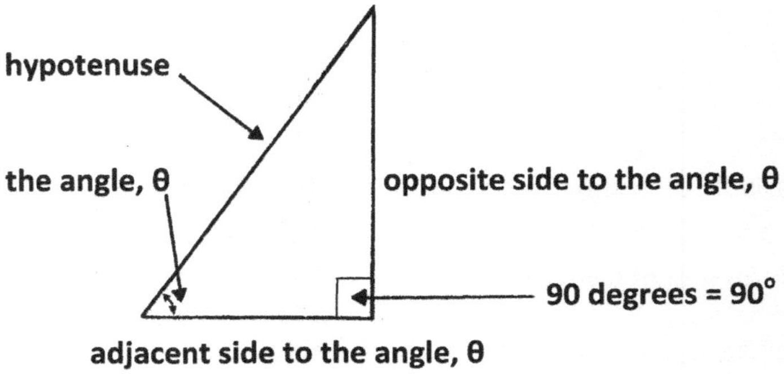

Sin - - Sine, a Trigonometry value: Sin θ = opposite side length/hypotenuse length

Cos - - Cosine, a Trigonometry value: Cos θ = adjacent side length/hypotenuse length

Tan - - Tangent, a Trigonometry value: Tan θ = opposite side length/adjacent side length

Sec - - Secant, a Trigonometry value = inverse or reciprocal of Cosine = Cos^{-1}; or; Sec θ = hypotenuse length/adjacent side length

Csc - -Cosecant, a Trigonometry value = inverse or reciprocal of Sine = Sin^{-1}; or; Csc θ = hypotenuse length/opposite side length

Cot - - Cotangent, a Trigonometry value = inverse or reciprocal of Tangent = Tan^{-1}; or; Cot θ = adjacent side length/opposite side length

Def - - Definition of a term or variable that defines what is meant by such a term or variable so that symbolism or communication is clear

e - - Base of Natural or Napierian logarithms, a constant irrational number which is approximately = 2.71828 . . . "e" may be used in an expression, such as: $y = e^x$

ln - - Natural or Napierian logarithm of base e – A natural logarithm of a number is the exponent or power that "e" is raised to in order to result in the desired number. For example: e times itself equals: e • e = e^2 and therefore ln e^2 = 2. For another example, e • e equals 2.71828 . . . multiplied by 2.71828 . . . equals 7.38904 . . . . Therefore: ln e^2 = ln 7.38904 . . . = 2.

anti-ln or antiln - - Anti-Natural or Anti-Napierian logarithm; reverse operation from that of a ln - From the above example: because ln 7.38904 . . . = 2 - Therefore: antiln 2 = 7.38904 . . .

log - - Logarithm, usually of base 10, equals the exponent power that the number 10 is raised to in order to result in a desired number – For example: log 100 = 2, that is another way of writing the following: 10^2 = 100 = 10 x 10; that is, to obtain the desired 100 number, the base number 10 is raised to, or multiplied by itself two times.

anti-log or antilog - - Anti-logarithm, usually of base 10; reverse operation from that of evaluating a log or logarithm – For example: antilog 2 = 100

$A_1 + A_2 + A_3 \ldots A_N$ - - Summation (or addition) of a series of numbers A_i to A_N. The symbol "N" used as a subscript in the series is the indicator of which A is meant. That is, A_1 means the first term in the series. A_3 is the third term in the series. A_N is the Nth term in the series of numbers.

A_1 , A_2 , A_3 , . . . $A_{N\text{-}}$ - - Set of elements or numbers, wherein the set has N members, wherein the members are denoted by A_i where each different "i" designates a different member of the set - The symbol "N" used as a subscript in the set of elements or numbers is the indicator of which A is meant. That is, A_1 means the first element or number in the set. A_3 is the third element or number in the set. A_N is the Nth term in the set of elements or numbers.

Mathematicians use various combinations of the above common symbols to link numbers, variables, and constants to produce Mathematical expressions. The above is a brief summary of symbols. Many others are used in special situations or Mathematical expressions.

PART III: GLOSSARY: Definitions of Mathematical Words and Phrases with Drawings and Illustrations (Alphabetically Arranged)

Abscissa – The distance of a point that is measured from the y-axis to the point, where the distance is measured parallel to the x-axis. It may be a positive number, if measured in the positive direction from the y-axis; or; it may be a negative number, if measured in the negative direction from the y-axis. In the below diagram the point is one unit to the right from the y-axis and therefore the abscissa of the point is "1" or "+1". The coordinate number "2" is 2 units up from the x-axis (parallel to y-axis) and is called the ordinate, valued at "2" for the below point.

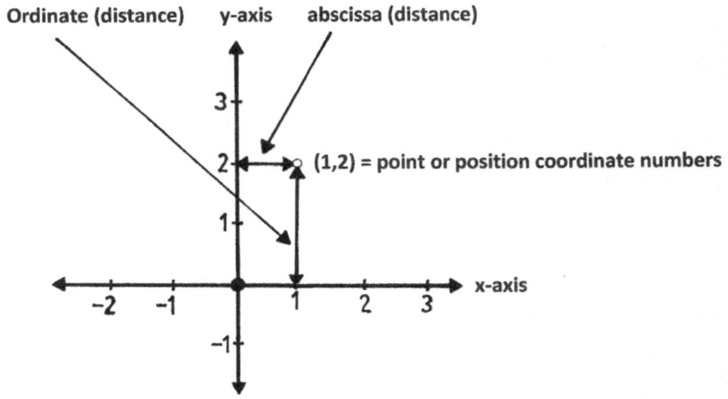

** Note: When defining the location or placement of a point, such as "o" above, on a coordinate system such as used above, the abscissa is always written before the ordinate with a comma between them and within parentheses, such as "(1,2)" above.

Absolute Value – Always a positive number value: The absolute value of a number is the numerical distance of that number from zero. The absolute value of a positive number is the same positive number. The absolute value of a negative number is the number but without the negative sign, therefore also making the absolute value of the negative number be positive. A vertical line both before and after a number or mathematical expression symbolizes or indicates the absolute value of the number or expression between the vertical lines. For example, the absolute value of 2 is written as |2|. Evaluating this gives the following: |2| = +2, or simply 2. The plus "+" sign is understood for a positive number without having the "+" sign there. An example of finding the absolute value of a negative number is: |- 4| = 4.

Account – A record of numbers, usually involving money.

Acute Angle – An angle that is less than 90 degrees (90 degrees is called a right angle.). Another way of writing "90 degrees" is "90°". The below shown acute angle is approximately 30 degrees, which is about 1/3 of 90 degrees.

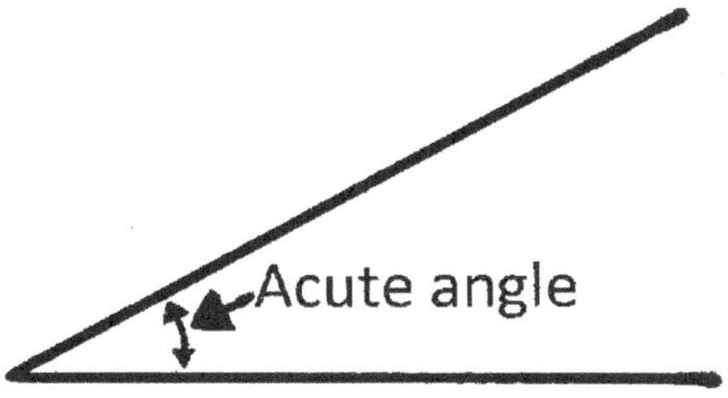

Acute Triangle – A triangle containing three acute angles, that is, the angles at all corners within the triangle are acute angles. The angles shown in the below Acute Triangle are all 60 degrees in angular value.

Addend – Any number being added to something else

Addition – The operation of combining two or more numbers by adding them to a get a resulting number called a sum: For example: 2 + 4 = 6 (6 is the sum of adding 2 and 4) – Another term for Addition is Summation

Additive Identity of Zero – Adding zero or "0" to any number does not change the value of the number. Therefore, N + 0 = N and also 0 + N = N

Additive Inverse – Adding the negative of any number to the number equals "0". For any number symbolized by N: N + (-N) = 0 or (-N) + N = 0 or –N + N = 0 (zero)

Algebra – A system for handling mathematical symbols and expressions, usually including letters and numbers that express desired arithmetic or mathematical operations and relationships

Algebraic expression – A group or series of numbers and/or symbols and/or variables that express an operation or series of operations – For example, Ax + By + 2

Algorithm – (a) A series of steps for calculating an answer, or, (b) any systematic method of solving a certain type of problem

Altitude or Height – (a) The length of a line that is perpendicular to one face of a polygon, figure, or shape that extends to an opposite vertex of the polygon, shape, or figure; or; (b) an elevation or distance from a level or horizontal plane; or; (c) the distance from top to bottom of something

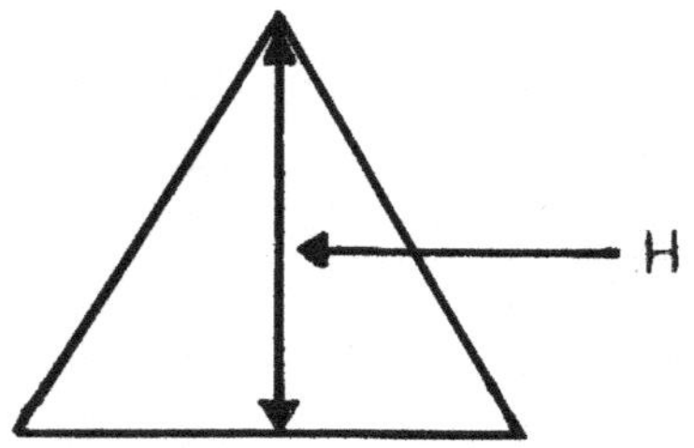

H is the Altitude of Height of the above triangle

Angle – The shape made by two straight lines meeting at a common point, usually measured in degrees (^{0}) or radians

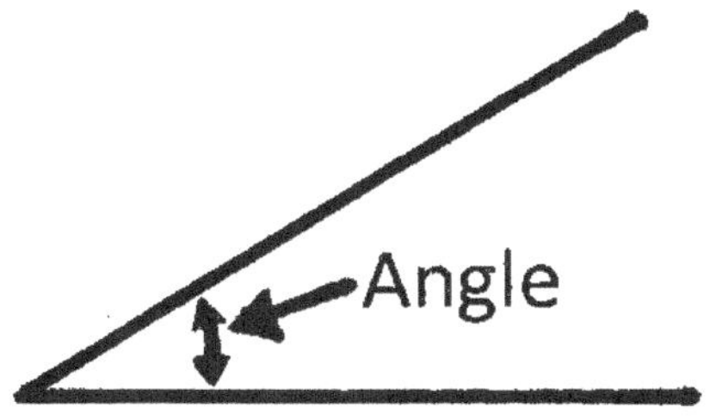

Angle of depression of elevation – An Angle between a horizontal line and a line of sight (down for depression and up for elevation)

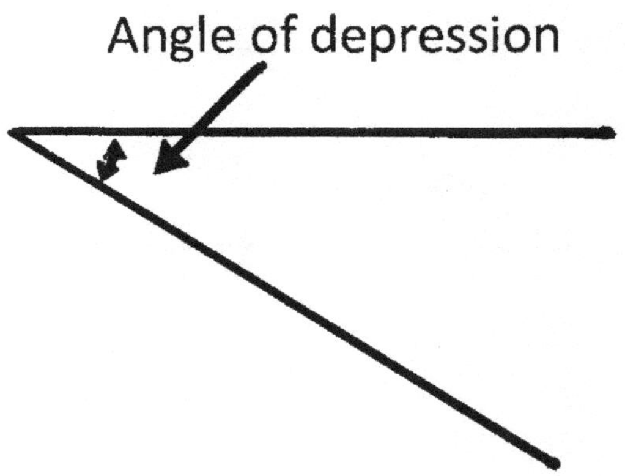

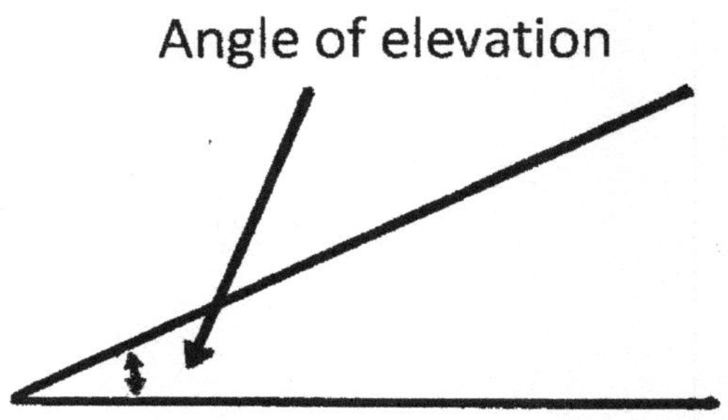

Apex – The point on a geometric figure farthest away from the base, base line, or base plane; usually the top or peak

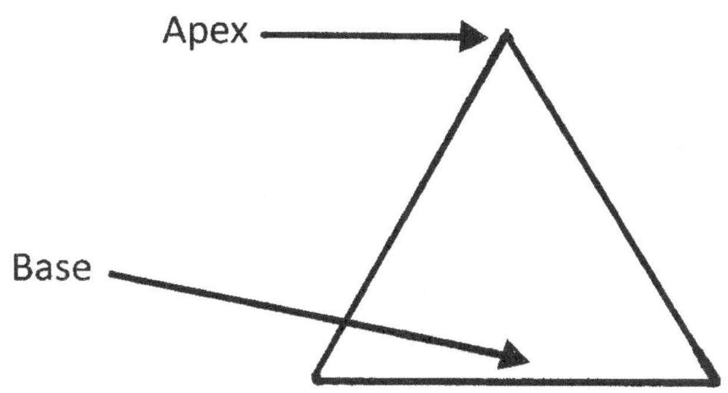

Apothem – A line perpendicular to a side of a polygon or figure that goes from the side to the center of the polygon or figure

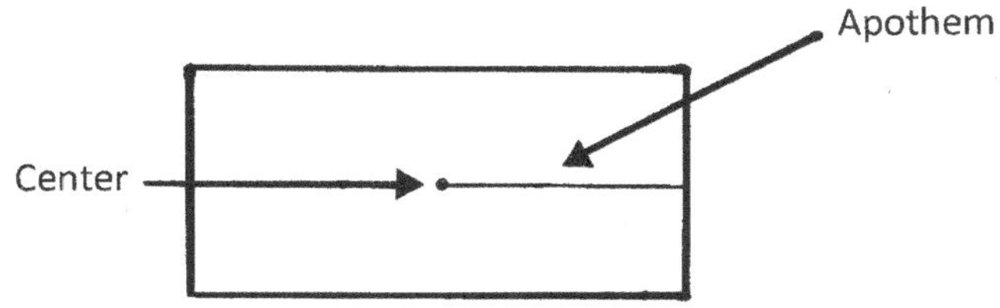

Approximate or Approximation – An estimate, guess, or mathematical result that is approximately correct or that is close enough to exactness to be useful for a particular purpose

Arc – Any part of a curved line, usually part of a circle, with two endpoints

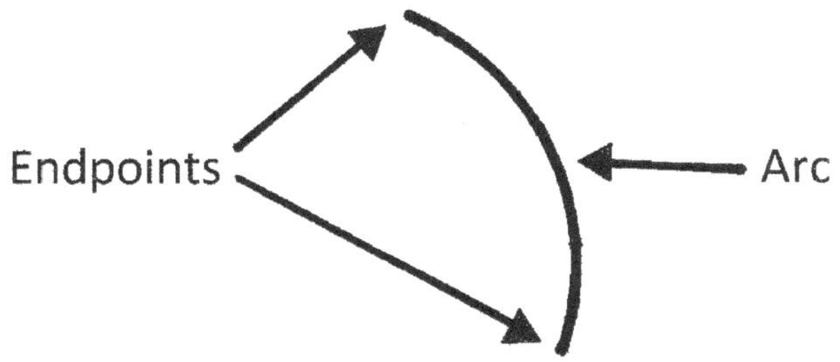

Area – A surface or region inside of the boundary formed by a figure, usually a surface or region within a planar figure, such as within a circle, square, rectangle, or other type of figure; however, an area can be the outer surface of any object having an outer surface – measured in square units – For example, the area of a room (12 feet long and 10 feet wide) in a house is 120 square feet, or written sometimes 120 ft^2.

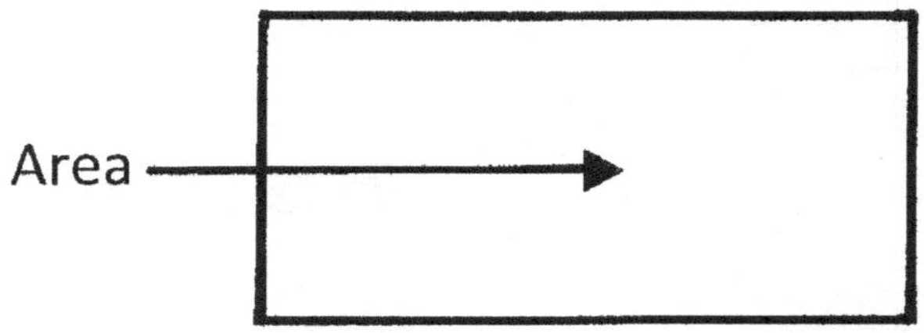

Argument – (a) An independent variable or section in a mathematical expression; or; (b) A sequence of propositions (propositions or premises) leading to a proposition called a conclusion

Arithmetic – The science or art of computing by adding, subtracting, multiplying, or dividing numbers

Arithmetic mean – The result obtained by adding together a series of numbers and then dividing by the number of numbers in the series – For example, the average or mean of 6 + 5 + 8 + 9 (note 4 numbers in the series) equals 28/4 = 7. "7" is the arithmetic mean (sometimes, simply called the average or mean) of the series: 6 + 5 + 8 + 9.

Arithmetic progression – A series of numbers in which the difference between consecutive numbers is the same as between other consecutive numbers in the series – For example, this is an Arithmetic progression: 1, 3, 5, 7, 9 with the difference between consecutive number being "2".

Arithmetic series – The sum of numbers in an arithmetic progression – For example, this is an arithmetic series: 1 + 3 + 5 + 7 = 16. The arithmetic series sums to a total of 16.

Associative Property of Addition or Associativity of Addition – The property of the operation of addition in which the result is the same regardless of the way the elements therein are grouped – For example for addition, 2 + (4 + 3) equals (2 + 4) + 3.

Associative Property of Multiplication or Associativity of Multiplication – The property of the operation of multiplication in which the result is the same regardless of the way the elements therein are grouped – Generally a property of addition and multiplication - For example, 2 x (4 x 3) equals (2 x 4) x 3.

Average or Mean – The result obtained by adding together a series of numbers and then dividing by the number of numbers in the series – For example, the average or mean of 6 + 5 + 8 + 9 (note 4 numbers in the series) equals 28/4 = 7. "7" is the average or mean of the series: 6 + 5 + 8 + 9.

Average deviation – The average of the absolute values of all of the deviations from the average in a set of data or numbers – For example, for the above series 6 + 5 + 8 + 9 with the average being "7", the deviations of each number in the series from the average of "7" are |6 - 7|, |5 - 7|, |8 - 7|, |9 - 7| which evaluating the absolute values give: 1, 2, 1, and 2 (4 deviations). The average of these deviations is (1 + 2 + 1 + 2)/4 which equals 6/4 which equals 3/2 or 1 1/2, or in decimal form 1.5.

Axiom – A statement or proposition that needs no proof because its truth is obvious, or, a statement or proposition that is accepted as true without proof; sometimes the starting point for further reasoning

Axis – A straight line through the center of a plane, plane figure, or solid, or, a straight line around which parts or points are measured or referenced relative to – See the x-axis and the y-axis shown above in the explanation of "abscissa".

Bar Graph – A graph in which the lengths of parallel bars are used to compare quantities – See below example where items A – F have related quantities ranging from about "1" to "6.5":

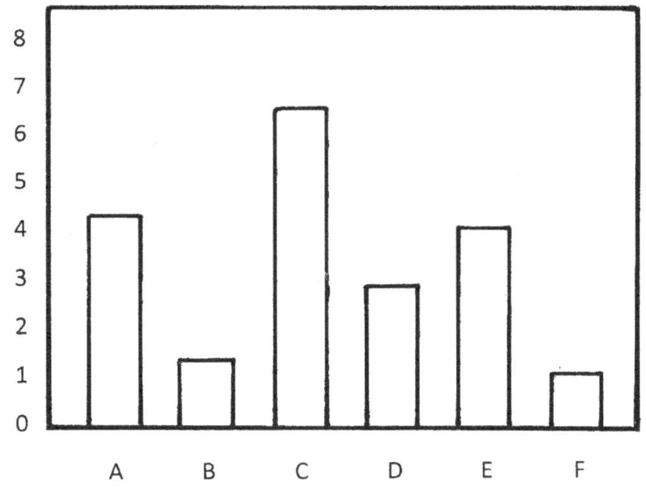

Base – (a) A number that is raised by the amount of an exponent, such as, "2" is the base in the number 2^3, wherein "3" is the exponent, or (b) The basic number value in a number system: For example, the Arabic number system uses 10 number symbols and is called a base 10 system.

Base of a triangle – The side that contains one end of the height or altitude of the triangle – See the above indication of a "base" of a triangle in the explanation drawing for Apex.

Bi-Implication or Biconditional – A logical connective between two statements, such as the expression stated as "If . . . ; then . . ."

Binary Number – A Number expressed in powers of two, using two digits, 0 or 1

Binomial – An algebraic expression with two elements or terms – For example: Ax + By

Bisect – To cut or divide something into two equal parts

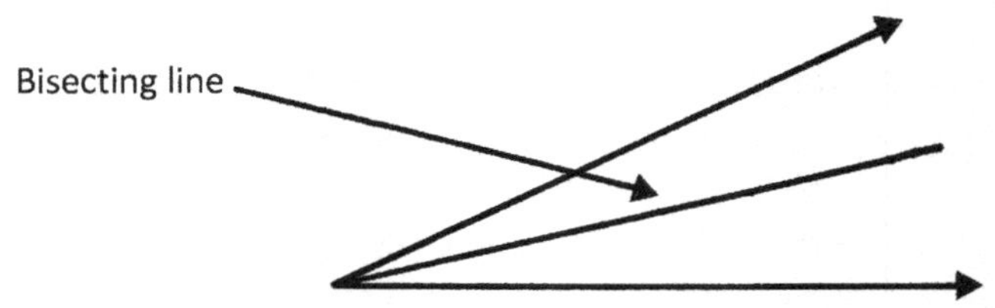

Bound – An item or element that can be only within a boundary or limit(s) in value

Boundary – A thing or line or figure that marks a limit, bound, boundary, or border

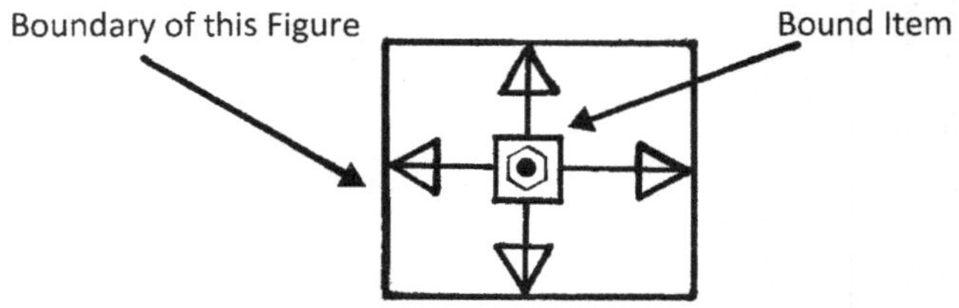

Calculate – Determine a result, compute a result, or figure out a result

Calculus – (a) The mathematics of change using the help and consideration of infinitesimals, or, (b) Any system of calculations using special symbolic notations

Cancel or Cancellation – To remove a common (same valued) item or common (same valued) factor from either both the numerator and denominator of a fraction or from both sides of an equation – For example, cancelling "2" from the numerator and denominator of the following fraction:

$$\frac{4}{10} = \frac{\cancel{2} \text{ X } 2}{\cancel{2} \text{ X } 5} = \frac{2}{5}$$

Central angle – An angle that has the center of a circle or sphere as its vertex

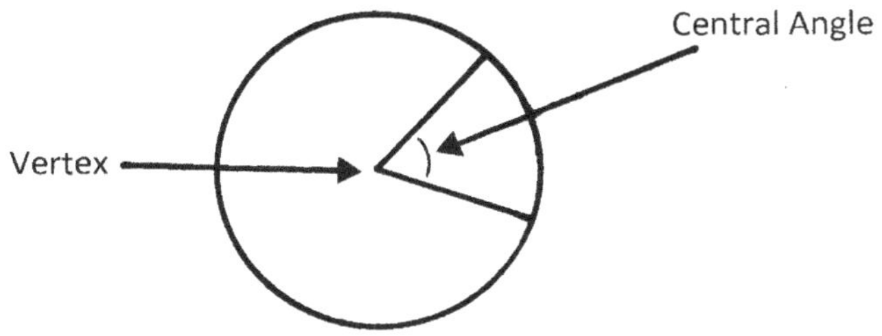

Cartesian Coordinate System – A two or three dimensional coordinate system wherein the coordinates are defined by two or three perpendicular axes, respectively, sometimes called a rectangular coordinate system. Points in space are defined in such coordinate systems by two or three numbers, usually relative to the axes: x-axis and y-axis (two-dimensional); or the x-axis, y-axis, and z-axis (three dimensional). See a Cartesian Coordinate System drawn in the above description of "abscissa" which is shown as a 2-dimensional coordinate system.

Cartesian Coordinates – For a two dimensional space, Cartesian coordinates are a pair of numbers that locate a point in a plane by its distances from two fixed, intersecting, perpendicular lines in the same plane. For example, in an x, y plane on which the fixed intersecting lines are usually designated as the x axis and the y axis; the pair of numbers would be the abscissa which is the distance from the point to the y axis and the ordinate is the distance from the point to the x axis. The abscissa and ordinate may be either positive or negative depending on which side of the corresponding axis that the point is. Cartesian coordinates may also be defined for a three dimensional space, usually with x, y, and z axes similarly to the two dimensional system. See a Cartesian Coordinate System drawn in the above description of "abscissa".

<u>Center</u> – (a) A point that is the same distance from all points on the circumference of a circle, or on the surface of a sphere, or, (b) a point that is the same distance from the vertices of a regular polygon

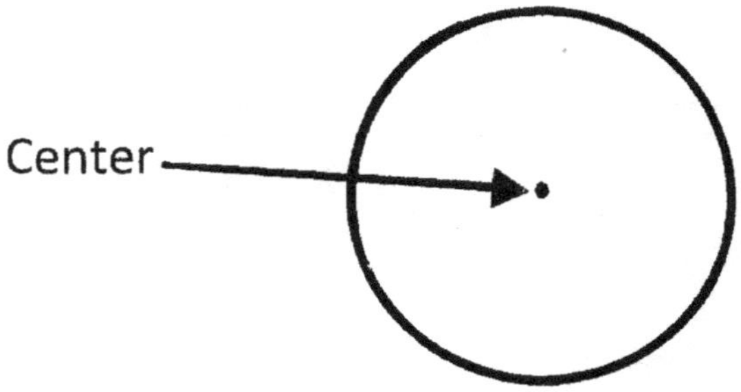

<u>Centi</u> – A metric prefix meaning a hundredth or 10^{-2} – For example, a centimeter means 1/100 of a meter, or, 0.01 meter.

<u>Characteristic</u> – (a) The whole number part of a logarithm value, or, (b) a descriptor word of phrase of something

<u>Chord of a circle</u> – A straight line segment joining two points on a circle, curve, or arc

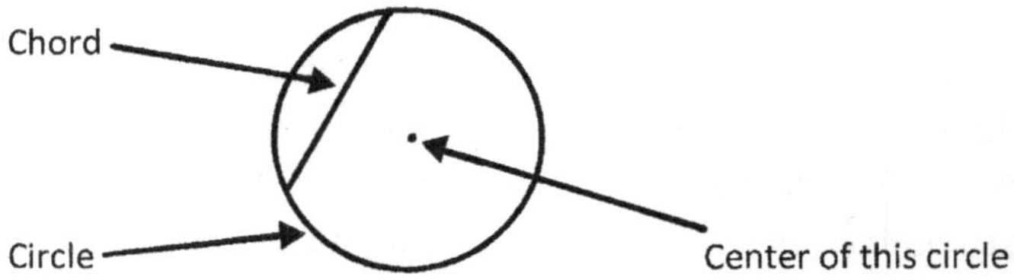

<u>Circle</u> – A plane figure bounded by a single curved line of which every point thereon is equidistant from a point at the center of the figure – See above circle.

Circumference – The length value or measurement of the distance around a circle or circular figure – The circumference of a sphere is generally the largest circle that can be formed on the surface of the sphere

Closed Form Problem – A problem that can be solved with an exact answer

Code – A symbolic form, or, to put in the form of symbols

Coefficient – A number or algebraic symbol prefixed as a multiplier to a variable or other quantity – For example, 5 is the coefficient in the Mathematical term 5x.

Cofactor – Any one of two or more factors or elements, the combined effects of which may cause a certain result or condition

Collinear – On the same line – For example, the following two arrow heads are collinear.

Combination – Any of a number of groupings or sets or subsets where the order does not matter – For example: 2, 4, 8 is the same combination as 4, 8, 2

Common Factor – A factor that is a factor of two or more numbers or terms – For example, the number "2" is a common factor for the following three numbers: 4, 6, and 10 because representing these numbers via factoring them, or breaking them up into the prime numbers that multiplied together result in each number, gives 4 = 2 x 2, 6 = 2 x 3, and 10 = 2 x 5

Common Fraction or Simple Fraction – A fraction whose numerator and denominator are both whole numbers – For example, 3/7

Common logarithm – A logarithm or exponent based on the number 10 – For example, the base 10 logarithm of "2" is written log 2. The numeric value of log 2 equals 0.301 . . . .

Commutative Property of Addition or Commutativity of Addition– The property of Addition in which the operation of addition wherein the order of elements or changing the order of elements does not affect the result – For example, 2 + 4 equals 4 + 2; that is, 2 + 4 = 4 + 2 = 6

Commutative Property of Multiplication or Commutativity of Multiplication – The property of Multiplication in which the operation of Multiplication wherein the order of elements or changing the order of elements does not affect the result – For example, 2 x 4 equals 4 x 2; that is 2 x 4 = 4 x 2 = 8

Compass (Mathematical or Geometry Compass)– An instrument consisting of two pointed legs that are connected by a pivot, used for drawing arcs or circles – Usually one leg is a metal point and the other leg is a pencil or pen point for drawing the arcs or circles. When drawing a circle, the metal point is placed on the desired center of the circle or arc so that the pencil or pen leg draws the circle or arc. The distance between the legs is adjusted to produce the desired Radius of the Circle. [This is different from a Magnetic Compass used for Navigation which points or indicates the direction to Magnetic North as well as directions of zero to 360 degrees.]

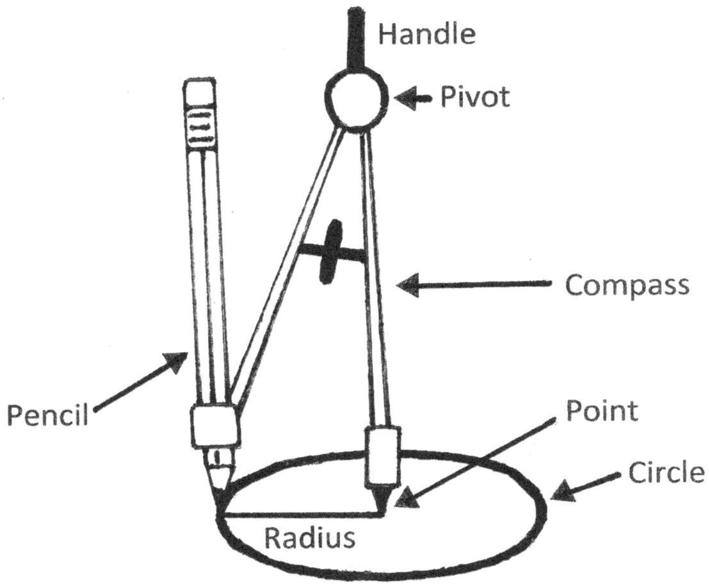

Compatibility Law of Addition – For any real numbers; a, b, and c: if $a < b$; then $a + c < b + c$

Compatibility Law of Multiplication – For any real numbers; a, b, and c: if $c > 0$ and if $a < b$;

then $a \times c < b \times c$

Complement – (a) The number or amount needed to complete something or bring it to completion, or, (b) either of two parts that make up a whole, or, (3) in angles the number of degrees that must be added to a given angle or arc to make it equal to 90 degrees or 90^0 (a right angle)

<u>Complementary angles</u> – Two angles that add in degrees to equal 90 degrees or 90^0 (a right angle)

Angle B is the complementary angle to Angle A as shown below:

Angle A plus Angle B equals 90 degrees

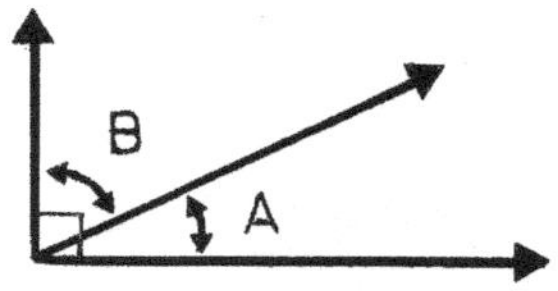

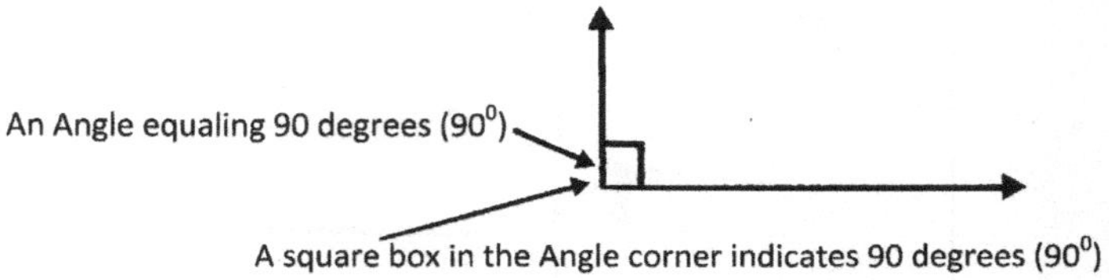

<u>Complex or Compound Fraction</u> –A fraction that contains a fraction in either its numerator or denominator or both – For example, the following are complex or compound fractions:

$$\frac{3/8}{5} \qquad \frac{7}{6/11} \qquad \frac{3 + 5/6}{7/12}$$

<u>Complex Number</u> – A general number type, usually expressed in the form: "a + bi"; where "a" and "b" are real numbers and "i" is the imaginary number which is the square root of -1 (symbolized as √-1).

Composite Figure – A figure or shape made up of two or more different figures or shapes, usually attached to each other – For example, a composite figure of a rectangle and a square is shown here:

Composite Solid – A solid made up of a combination of two or more other solids, usually attached to each other – For example, a composite solid of a parallelepiped and a cylinder is shown here:

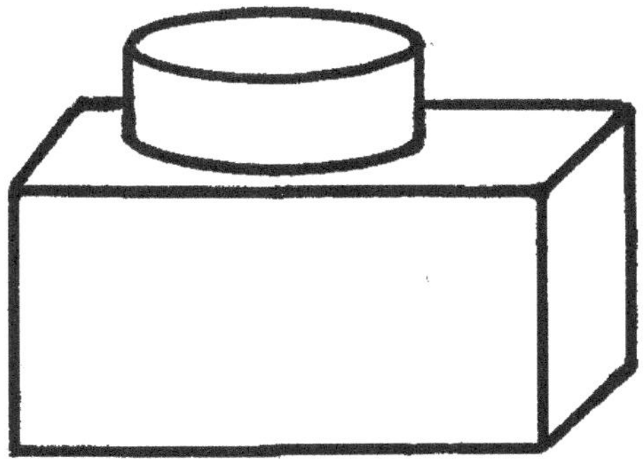

Concave polygon – A polygon with one or more diagonals between vertices that have points outside of the polygon interior – For example; see the diagonal that is drawn "outside" of the Hexagon interior:

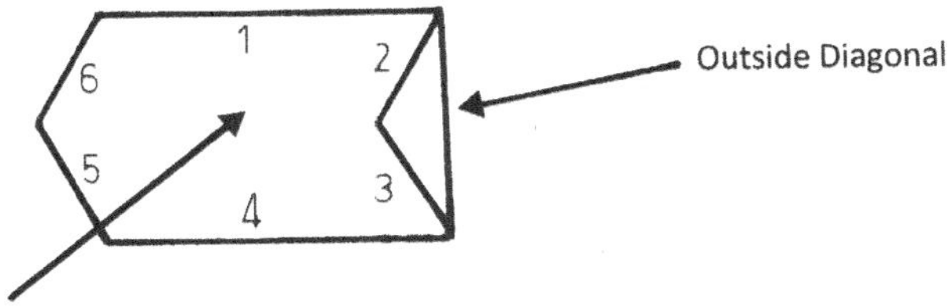

Conic Section – The figure produced by the intersection of a cone with a plane – Examples of conic sections are: parabola, hyperbola, ellipse, and circle.

Conclusion – (a) The ending or result of a Mathematical computation or process; or; (b) The last step in a reasoning process or series of steps or operations; or; The final proposition that follows from a series of propositions (premises or hypotheses)

Cone – A three-dimensional figure with one curved surface, one flat surface (usually circular), and one vertex – For example, a cone is drawn here:

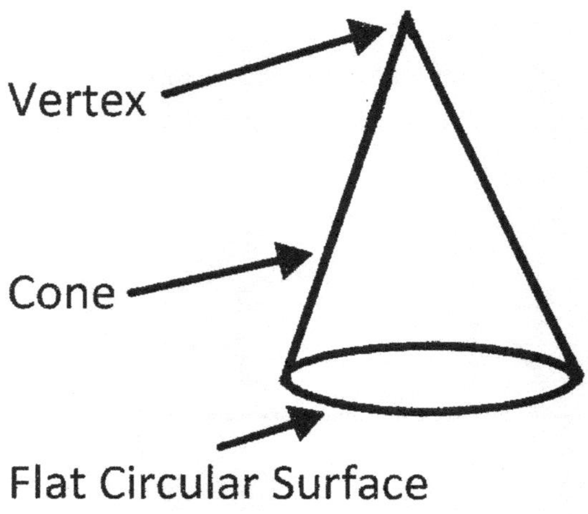

Congruent – The state or property of being in agreement; or; the property of a figure, plane, or solid figure whereby it coincided with another figure, plane, or solid figure, after it is moved, rotated, or flipped over – For example, drawn here are three congruent triangles.

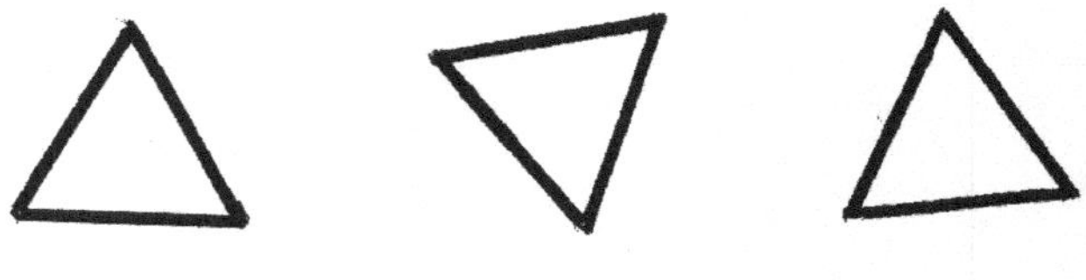

Conjugate angles – Two angles that add or sum together to result in 360 degrees - For example, drawn below are Angle A and Angle B which add together to result in a full circle of 360 degrees (360^0)

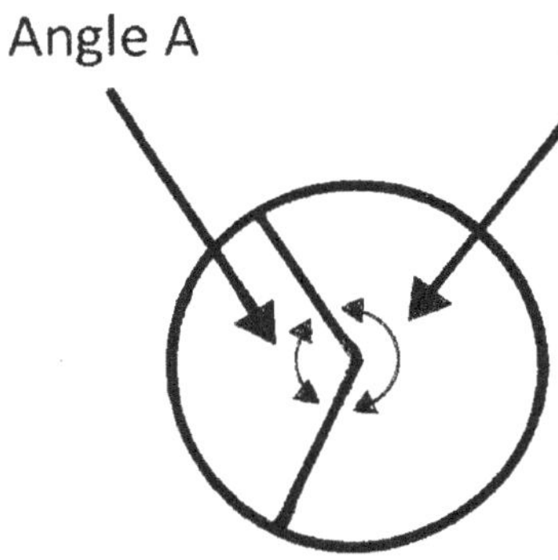

Connective or Conjunction – Joining two items or elements together – For example, the logical connective "and" joins together two items as shown here: A and B

Constant – Anything that does not change or vary in amount, value, or character

Coplanar – In the same plane

Consecutive – Lined up in order one after another in some sequence – For example, here are some consecutive numbers: 1, 2, 3, 4, 5

Conversion of units – Changing the numerical representation of something to another different unit designation. For example, a conversion of a length would be done by starting with a length of 6 feet and changing the units of feet to yards. The changed 6 feet length can then be converted to units of yards as 2 yards. The conversion keeps or maintains the same physical length but represents the length now in yards instead of feet.

Convex polygon – A polygon with all interior angles measuring less than 180 degrees - For example, a square is a convex polygon:

[In a square all internal angles are 90 degrees (90^0)]

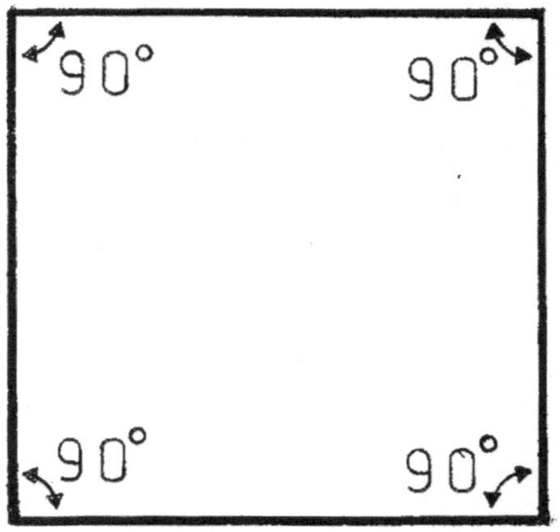

Coordinate – Any item of a set of numbers defined within a coordinate system that locates the position of an item or thing or point, the position usually being defined as two or more numbers or variables. See the abscissa and ordinate shown in the above definition of abscissa.

Coordinate axes – Paired perpendicular number lines on a two-dimensional Cartesian coordinate system or plane, or three(3) perpendicular number lines in a three-dimensional Cartesian coordinate system or 3-dimensional space. See 2-dimensional coordinate axes in the above definition of abscissa.

Coordinate system or grid – A system or grid with at least two axes that is useful to locate the position of one or more points therein – See a coordinate system or grid in the above definition of abscissa.

Coordinate plane – A plane containing two of the coordinate axes of a coordinate system

Coplanar – Located within the same plane

Criteria – Plural form of criterion, sometimes also used as a single criterion in context

Criterion – A standard or rule by which something can be judged or evaluated, or, a measure of value

Cross section – A shape or figure formed when a plane cuts through a three-dimensional figure - For example, see a circle produced when a plane cuts perpendicularly through a tubular cylinder

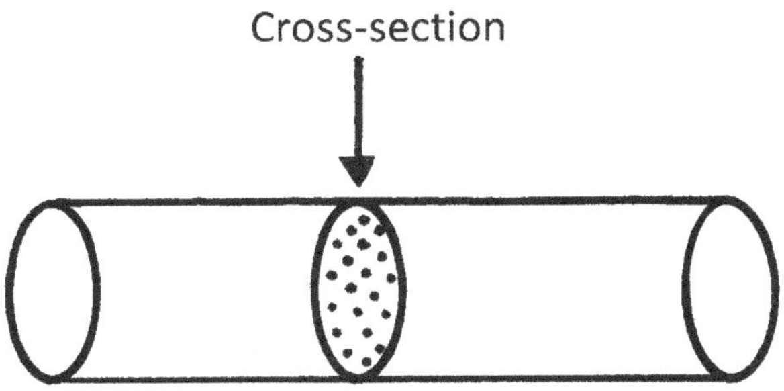

Cube – A solid shape of solid figure with six congruent or identical square faces

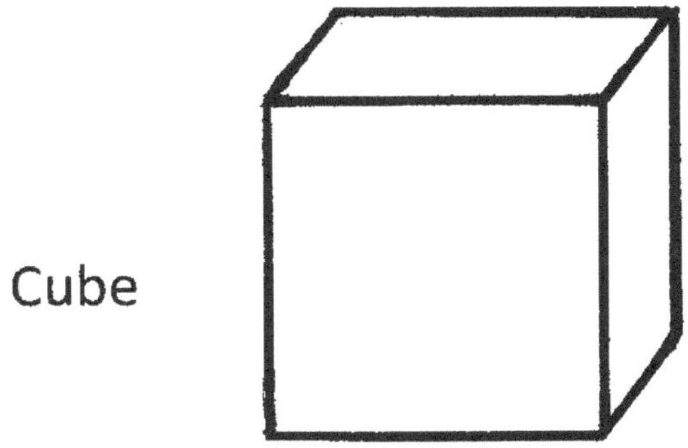

Cube of a number – A number raised via the exponent "3" to what is termed the third power or termed cubed. For example, 2 cubed is designated 2^3 which is 2 times 2 times 2 which equals 8. Note that the third power is obtained by using the base three times via two multiplications. The base is the number that is being cubed, "2" in the above example.

Cube root – A number, called a cube root, which when cubed, or raised via the exponent "3", results in the number for which the cube root is defined or desired. For example, the cube root of 8 is designated $\sqrt[3]{8}$ and has the value "2". That is, the cube root, which is "2", when cubed, means 2 times 2 times 2. This cube root of 8 being "2", when multiplied twice by itself, or cubed (2 times 2 times 2), results in the number 8. As an equality expression, this is expressed as $\sqrt[3]{8} = 2$. Alternatively, sometimes a root of a number is expressed using exponential expression which would be written for the cube root of 8 written as $8^{1/3} = 2$.

Cubic unit – A measurement unit used for expressing a three-dimensional measurement, such as a volume. For example, a volume may be expressed in cubic feet. A large ice cube with each side being of length of 1 foot has a volume of 1 foot times 1 foot times 1 foot which is 1 cubic foot, also designated as 1 ft^3. Non-cubic shaped items may also have a volume in cubic feet. For example a water heater tank in a house is cylindrical in shape and may have a volume of 40 cubic feet (40 ft^3).

Curve – A line having no straight part

Curve

Cycle – (a) A single event that is repeated in a series, or, (b) A repeating period of time, or (c) An event in a repeating series of events.

Cyclic – An item or element that repeats or recurs

Cylinder – (a) A solid figure consisting of two parallel identical bases that are joined by a smooth curved surface, or, (b) The commonly thought of cylinder is actually a special cylinder type which has a top and bottom plane and encloses a volume via right angles between the top, sides, and bottom and is called a right circular cylinder

Right Circular Cylinder 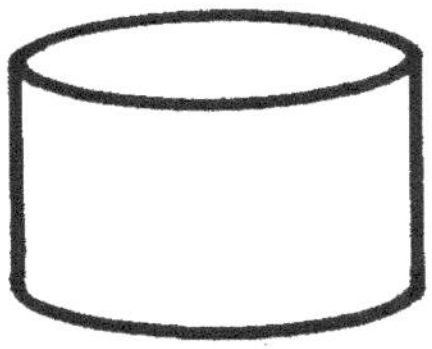

Cylindrical surface – a surface made up of a collection of a plurality of parallel lines that collectively form a surface – See above shape which may be seen as a collection of parallel lines arranged side by side around the circular outer vertical surface of the above cylinder.

Data – information of any kind

Deci – A metric prefix meaning a tenth or 10^{-1} – For example, a decimeter means 1/10 of a meter, or, 0.1 meter.

Decimal – (a) A fraction based on Arabic numerals with an unstated, but assumed, denominator of 10 or some power of 10, or, (b) A number based on the Arabic numbers ten (10) – For example, "0.436"

Decimal Fraction – The part of a Decimal number to the right of any integer number part, if present. Generally, a Decimal Point (period ".") separates the Decimal part of the number from the Integer part of the number. In the Decimal number 7.444, the Decimal Fraction is ".444" and the integer number part is "7". A Decimal Fraction without an integer number part may be ".36", for example. Usually a Decimal number or Decimal Fraction without having any value in an integer number part will place a zero "0" there as a place holder, or, for clarity, that there is no integer number value. For example, usually the Decimal Fraction ".36" will be written "0.36". The "0" is actually an integer number but; of course, it has no measurement value, being only zero "0".

Decimal Number – A number made up of, or containing, a decimal fraction part and maybe an integer number part. For example, the Decimal number "5.34" contains the decimal fraction part: ".34" and the integer number part: "5". Another Decimal number may be .88, which contains only the decimal fraction: ".88", frequently written with a "0" integer number place holder as "0.88".

Decimal Point – The period "." in a number just to the right of the lowest value whole number digit and just to the left of the first fractional digit. The digits to the right of the Decimal point "." have a value that is less than 1. A Decimal point is a place holder to separate the left-hand numbers from the right-hand numbers in a number in a Decimal Number. – For example, in the Decimal number: 45.13, the 4 and 5 designate the numerical values of 40 and 5, respectively. The 1 designates the numerical value of 1/10 and the 3 designates the numerical value of 3/100.

Decode – To translate a coded value or message into something easily understood

Definition – A statement of what something is, or, stating the meaning of a word, phrase, or any specified item

Degree – (a) A unit of measure for angles or arcs; which is 1/360th of a full circle, usually symbolized by "°", or, (b) for algebraic expression the rank of such an expression determined by summing the exponents of terms therein, or, (c) the largest power or exponent of a term in a polynomial expression, or, (d) the extent or amount of something

Denominator – The element or number below the line (called Fraction Bar) or diagonal in a fraction – For example, in the fraction ¾, the denominator is "4".

Dependent variable – A variable item whose value is determined by the value of at least one other variable, usually an independent variable or another dependent variable

Deviation – The difference in numerical value between one number and another – For example, the deviation of 5 from 3 is "2".

Diagonal – A line that is not a side of a polygon which also connects two different vertices in a polygon, figure, or shape

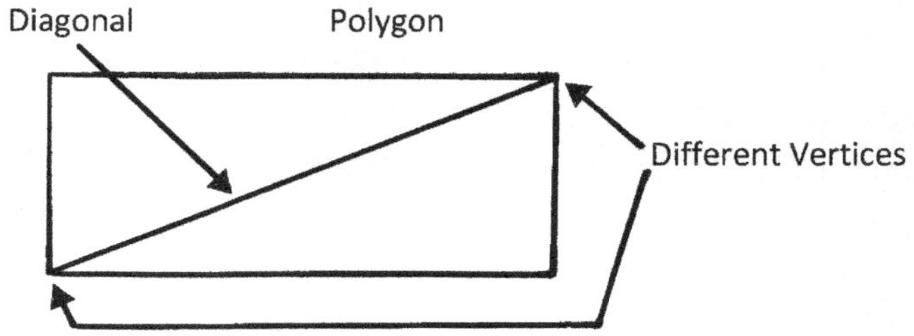

Diameter – The length of a straight line segment that lies within a circle or sphere that passes through the center of the circle or sphere and has endpoints on opposite points on the circle or sphere – A diameter of a circle or sphere is two radii in length

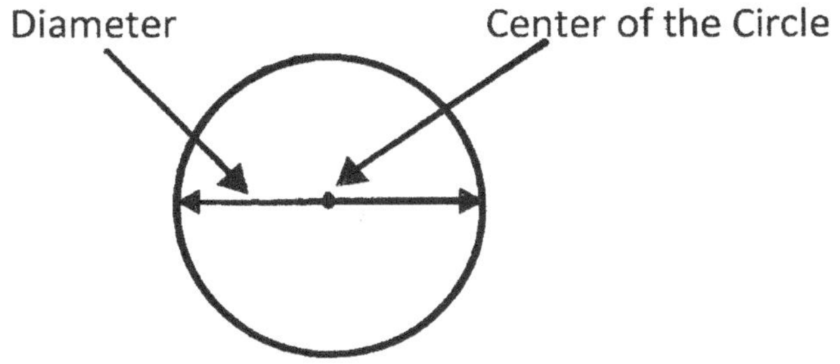

Discrete change – A specified amount of change of a quantity

Distance – The length of a gap, space, or interval between two points

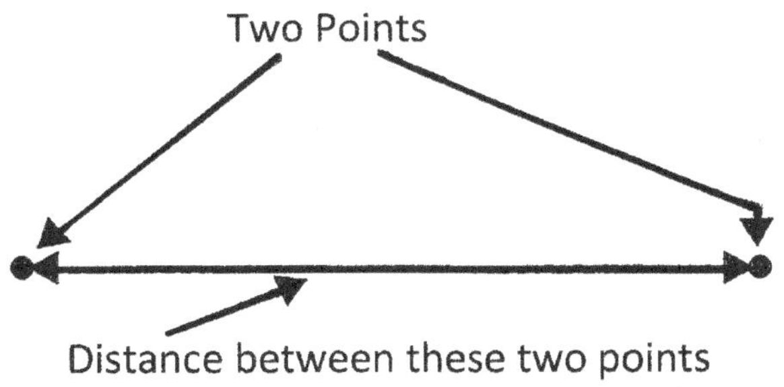

Distributive property – The property of addition combined with multiplication in which multiplication of a number, the multiplier, by an expression containing numbers added together equals the additive sum of each added number multiplied separately by the multiplier. For example, for a multiplier number, a, multiplied by the sum of b + c; this equals a multiplied by b plus a multiplied by c; shown in symbols as follows: a(b + c) = ab + ac or a x (b + c) equals (a x b) + (a x c) or a(b + c) = a • (b + c) = (a • b) + (a • c); or; another way of saying this is that a number that multiplies a sum of numbers equals the sum of each number of the original sum each multiplied by the multiplier. Yet another way of saying this is that a number that multiplies numbers added together is distributed by multiplication separately over each and all of the initially added together numbers. This property also applies for multiplying a longer series of added numbers. For example: a (b + c + d + e) = ab + ac + ad + ae.

Dividend – A number or quantity that is intended to be divided by a divisor to result in a quotient answer: For example, in the Division of 1000 by 5; 1000 is the Dividend and 5 is the Divisor; symbolized also as 1000/5. 1000 divided by 5 equals 200 as the answer. The 200 resultant answer in this division is called the Quotient.

Special case Note: Dividing by zero (0) gives an indeterminate result. For example: 5/0 is indeterminate, that is, cannot be determined.

Divisor – A number of quantity that divides another number or quantity called a dividend to result in a quotient (answer)

Domain – A set of independent variable values for which a function or relation can be evaluated or defined

Edge – The line segment where two faces of a solid figure or shape meet

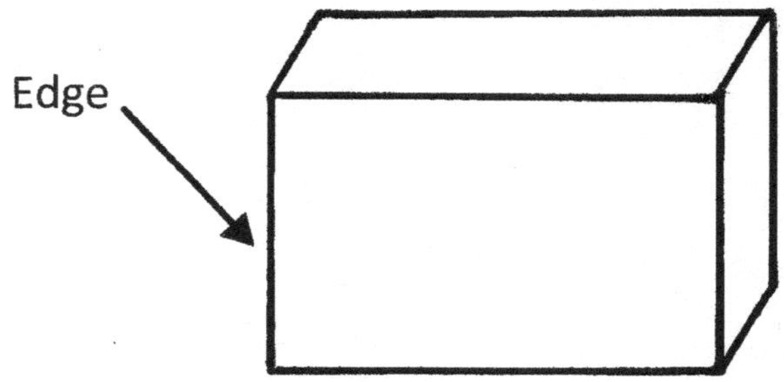

Ellipse – A closed planar figure wherein each point thereon satisfies the condition that the sum of the distances from two points, called foci, is the same for all points on the ellipse; An ellipse is a type of conic section.

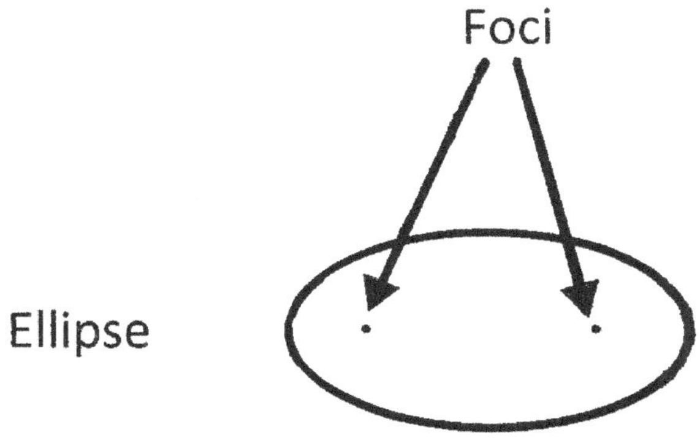

Endpoint – A point that marks the end of a line segment

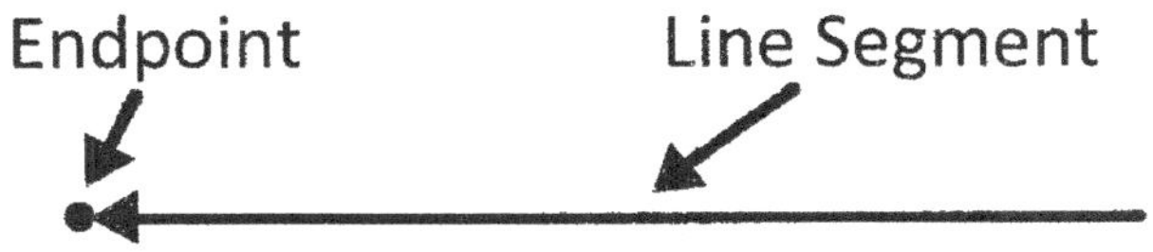

Equal – Two items or expressions having the same value – For example, in an algebra equation an equal sign "=" is used to indicate that one side of the equation equals the other side. In the equation written as "Y = 3X + 5" one side is "Y" is indicated in the equation as being "equal" to "3X + 5".

Equation – A statement of equality between two quantities or expressions, usually containing symbols - For example: y = 2x + 6 means that the variable "y" equals "2x + 6".

Equation of a line in an x-y plane – An equation usually of the general form: Ax + By = C, which describes a line present in an x-y plane

Equidistant – A set of points that are the same distance from a reference item, usually a point or line

Equilateral triangle – A triangle whose sides are all of the same length – See below example:

Error – (a) Something that is incorrect or wrong, or, (b) the difference between the true or correct value of an element or term and the value of that element or term that is measured or approximated

Error bar – A Graphical indication shown with a value or data point on a graph which designates or shows the expected or approximated upper and lower values within which the true, correct, or actual exact value may be; usually shown as a short line that passes through the value or data point with short perpendicular lines at its top and bottom (or right and left) showing the upper and lower possible values of the point being depicted near the center of the error bar lines

Enlarged example of a Data point example with error bars:

Estimate – To calculate approximately, or, to make an estimation

Estimation – An estimate of what value an element, term, or variable may actually be

Even number – (a) An integer number that is exactly divisible by two to result in another integer number – Examples: 2, 4, 6, 8, 10, 12, 14; or; Sometimes Even Numbers are defined as inclusive of positive integers or negative integers that are exactly divisible by two to result in another integer.

Exercise – Performance of an activity so as to improve or perfect a technical skill or understanding of problem solving

<u>Exponent</u> – A small figure or number generally placed above and to the right of a term or element to show how many times that the term or element is used as a multiplicative factor – an exponent may be positive, negative, fractional, zero, or even itself a variable in numerical value – For example, in the number "4^3", the "3" is an exponent.

<u>Exponential or Exponential form</u> – Relating to an exponent, or, an expression containing one or more exponent(s) or exponential number(s) – For example: $y = 5x^3$

<u>Expression</u> – A symbol or set of symbols that show one or more relationships between the symbols

<u>Exterior angle of a triangle</u> – Any angle outside of a triangle defined by one side and extending the adjacent side outside of the triangle

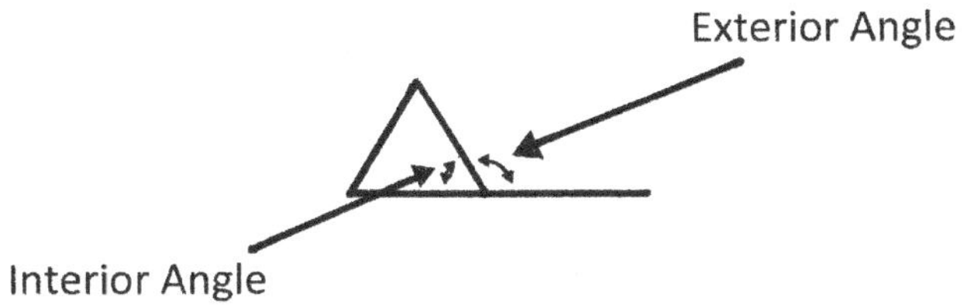

The sum of the degrees of an interior angle plus its corresponding exterior angle as diagrammed above equals 180 degrees. An interior angle is therefore a supplementary angle to its corresponding exterior angle. Also, the summation of all three of the exterior angles of a triangle is 360 degrees. The summation of all three interior angles of a triangle is 180 degrees.

<u>Face</u> – A plane figure or shape that is one side of a solid figure or shape

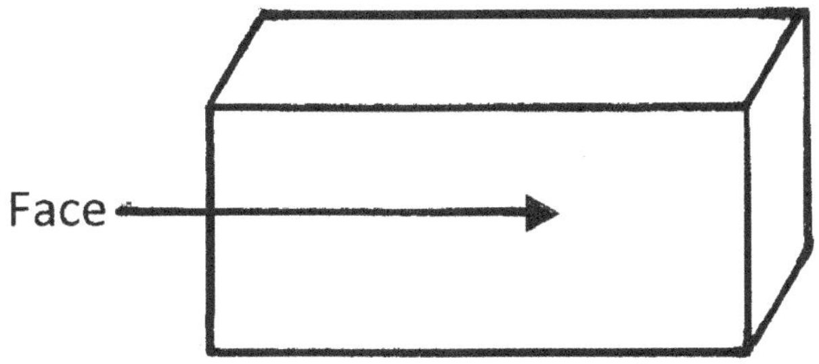

<u>Factor</u> – (a) A portion of a number or mathematical expression, usually a term (factor) that is multiplied by another term (factor)- For example, the number 6 can be factored into factors "3" and "2" because, using multiplication, 3 x 2 = 6, or, (b) the process of subdividing a complex expression into simpler parts (that is, factors), frequently utilized in subdividing a number or expression into quantities that when multiplied together result in the original number or expression

Factorials – A multiplicative product of a series of consecutive whole numbers, beginning with one (1) and ending with a specified number. For example, 4! (stated 4 factorial) is 1 X 2 X 3 X 4. Completing the multiplication results in the following: 4! = 24.

Femto – A metric prefix meaning a quadrillionth or 10^{-15} – For example, a femtometer means 1/1,000,000,000,000,000 of a meter, or, 0.000,000,000,000,001 meter.

Figure – (a) the outline or shape of something, or, (b) to calculate a mathematical answer

Finite – (a) anything that has measureable limits, or, (b) a term or element that is capable of being determined and is neither infinite nor infinitesimally small – For example: A finite number is the number "5".

Finite Series – a series containing a limited or countable number of terms – For example, a finite series is the following: y = 1 + 4 + 5 + 6

Flow Chart – A chart or diagram that shows relationships or the sequence or order between steps in a process, or, relationships between collections of items can be diagrammed in a flow chart like format

Formula – An expression that defines what is performed to produce a desired result

Fraction – (a) Breaking or dividing something into smaller parts, or, (b) A quotient of two numbers or two Math expressions with the numerator or dividend being the top number or expression and the divisor or denominator being the bottom number or expression

Fraction Bar - The line (called Fraction Bar) or diagonal in a fraction – For example, in the fraction ¾, the denominator is "4" which is separated from the numerator which is "3" by the Fraction Bar line.

Frequency – The number of times that an event or item is present in a time interval or some desired space

Function – An association between terms wherein at least one term is an independent variable and at least one term is a dependent variable. For example, f(x) = . . . (stated as a function of x equals . . .) – A function is formally defined as a relation in which each domain [input(s) or argument(s)] member or set or domain members is related or evaluated via the function to result in one and exactly one range (output value) member. Other examples of function notations are: g(x), f(x,y,z), H(a,b), etc.

Geometry – The branch of mathematics that deals with points, lines, planes, and figures; and examines their properties, measurement, and relationships in space

Giga – A metric prefix meaning a billion or 10^9 – For example, a billion dollars is 1,000,000,000 dollars

Googol – A metric prefix meaning 10^{100} – For example, a googol gram is 10^{100} grams

Graph – A diagram that may contain a curve, line, broken line, or a series of bars; representing various kinds of quantitative information and relationships – The below example, relates percentages to time in Days for two different conditions, one shown with closed circle points and the other shown with closed triangular points.

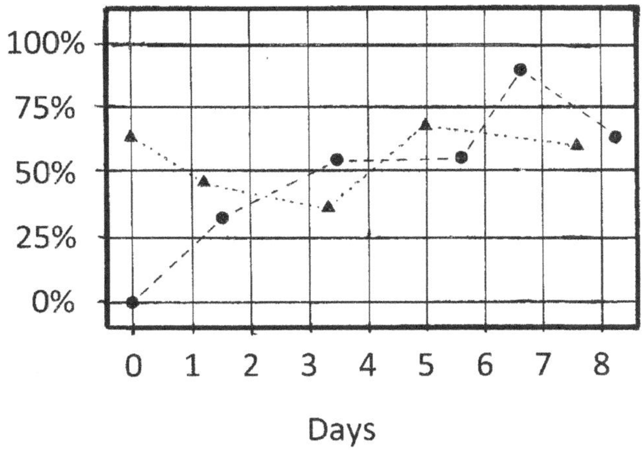

Grid – A pattern of lines usually forming squares and which are usually horizontal and vertical in a coordinate system – For example, see the below grid which contains 10 horizontal lines and 10 vertical lines:

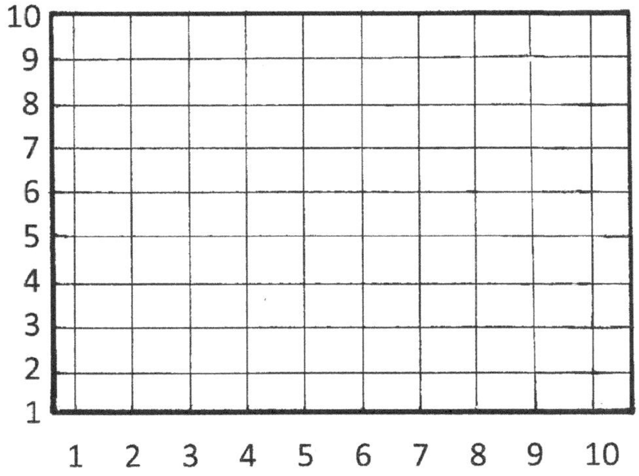

Half line – A line with one endpoint and which extends indefinitely in one direction from that endpoint

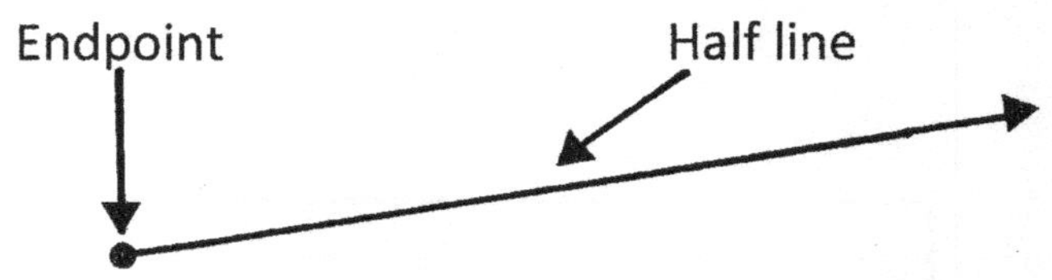

Half Space – Each space defined or separated from another Half Space by a plane which separates a three dimensional space into two parts, called half spaces

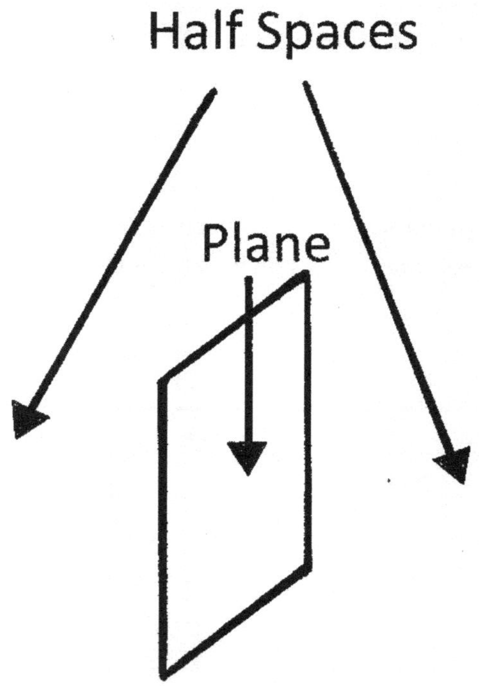

Height or Altitude – (a) The length of a line that is perpendicular to one face of a polygon, figure, or shape that extends to an opposite vertex of the polygon, shape, or figure, or, (b) an elevation or distance from a level or horizontal plane, or (3) the distance from top to bottom of something

H is the Altitude or Height of this triangle:

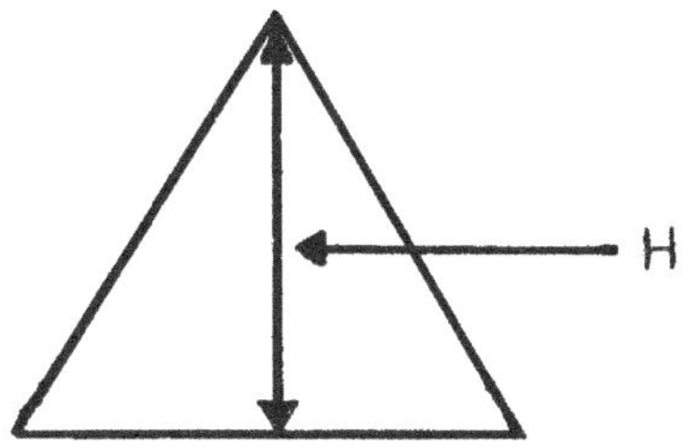

Hexadecimal Number – A number defined and written in a base 16 number system

Hexagon – A planar figure with six sides and six angles

Histogram – A bar graph in which the area of each bar is proportional to the relative amount indicated

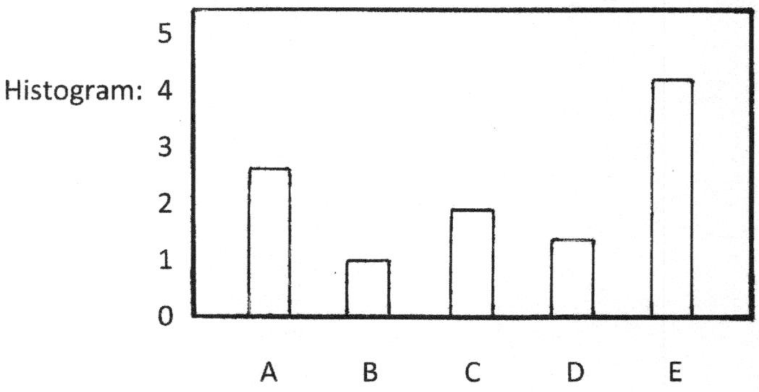

Hyperbola – A planar figure defined by points that are a constant difference in distance from two other points called foci – The foci are depicted below as the points, "F", and the solid lines are the hyperbola lines.

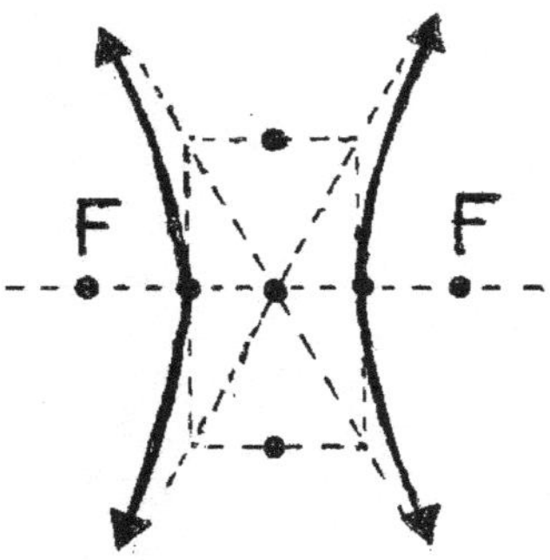

Hypotenuse – The longest side of a right triangle

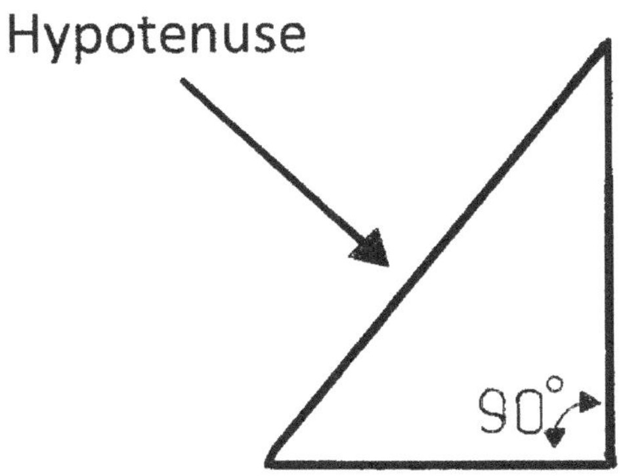

Hypothesis – (a) A basis for further investigation; or; (b) An unproved theory; or; (c) A tentatively or temporarily accepted proposition or proposal to explain certain facts; or; (d) A premise

Imaginary Number – An imaginary number is the product of a non-zero real number times the imaginary unit. The imaginary unit is the square root of negative one, symbolized as "i" which is the symbol for √-1. An imaginary number is symbolized as bi, wherein "b" is a non-zero real number and "i" is "√-1". Note that an imaginary number is a special type of complex number. A complex number is usually symbolized by the form "a + bi". "a" is a real number.

Improper Fraction – A fraction in which the denominator is less than the numerator in value – For example, the fraction: 7/5

Independent variable – A variable whose value may be determined freely without reference to other variables

Indeterminate form – A mathematical expression that cannot be evaluated in exact terms – For example, a fraction with zero (0) in the denominator such as: 5/0

Index – (a) A subscript, usually a number, for indicating specific items in a series of items, such as A_i wherein the "i" is the subscript index, or, (b)The exponent integer or symbol placed above and to the left to indicate a root number - For example, "3" is an index in the number $\sqrt[3]{8}$ which indicates that the cube root is meant.

Inequality – The relationship between two unequal quantities or terms – Some statements containing inequalities:

Description	Symbol
A is less than B	A < B
A is greater than B	A > B
A is less than or equal to B	A ≤ B

Infinite – (a) Lacking a limit or bound, or, (b) indefinitely large, or, (c) greater than any finite number, no matter how large – Example of something that is infinite: infinity

Infinitesimal – (a) Something too small to be measured, or, (b) A mathematical variable that approaches zero as a limit

Infinite wedge – Four wedge spaces are produced in three dimensional space by two intersecting planes, such as shown below: The depicted four wedges extend out infinitely.

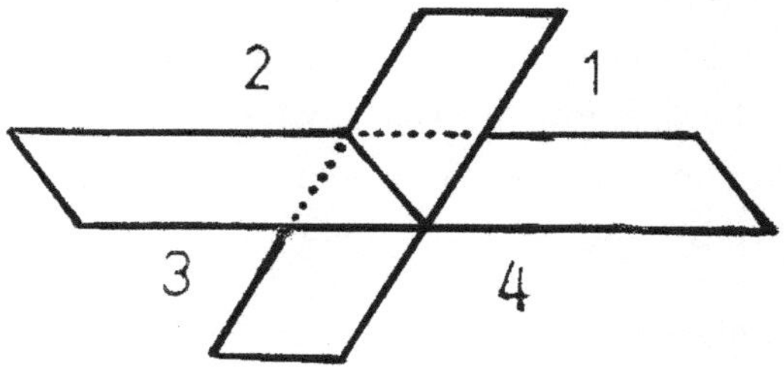

Instantaneous rate of change – A rate of change existing at a particular instant in time

Integer – Any positive or negative whole number or zero – Examples of integers: . . . -2, -1, 0, 1, 2, 3, . . .

Interior angle of a Triangle – Any angle at a corner of a triangle inside of the triangle

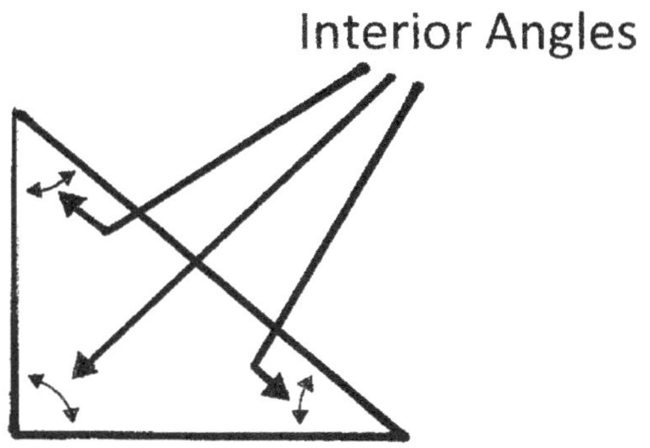

Exterior angles are related to Interior angles. An Exterior angle of a triangle – Any angle outside of a triangle defined by one side and extending the adjacent side outside of the triangle

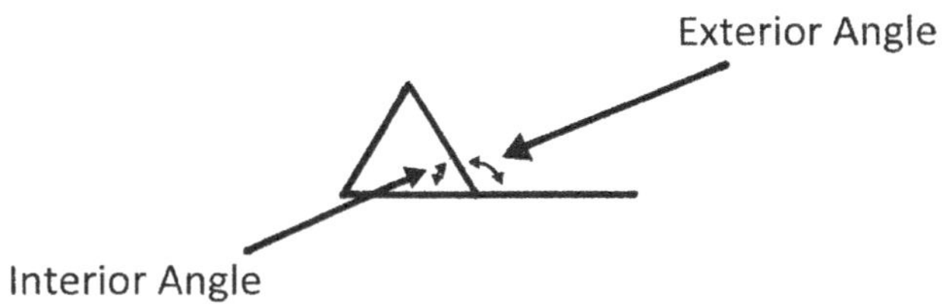

The sum of the degrees of an interior angle plus its corresponding exterior angle as diagrammed above equals 180 degrees. An interior angle is therefore a supplementary angle to its corresponding exterior angle. Also, the summation of all three of the exterior angles of a triangle is 360 degrees. The summation of all three interior angles of a triangle is 180 degrees.

Intersecting Lines – Lines that cross each other at a common point (A common point is a point that is the same point on both lines.)

Intersection – The set of points containing all of the points in common to two or more sets

For example: Given Set 1: 4, 6, 8, 9, 2, 5, 7; and Given Set 2: 3, 5, 6, 7, 9, 12; the Intersection set between Set 1 and Set 2 is: 6 and 9

<u>Irrational number</u> – A real number that is neither expressible as an integer nor the quotient of two integers – For example, the constant "π" equals 3.14159 . . . and is an irrational number. Also, a Non-Repeating and Non-Terminating Decimal Number is Irrational.

<u>Inverse</u> – (a) Reversed in order, or, (b) an operation when performed exactly cancels another operation, or, (c) in division by dividing 1 by the number as in 1/x which is the inverse of x

<u>Irreducible</u> – Cannot be reduced to something simpler

<u>Irregular polygon</u> – A polygon whose sides are not all of the same length

Isosceles Triangle – A triangle with at least two equal sides

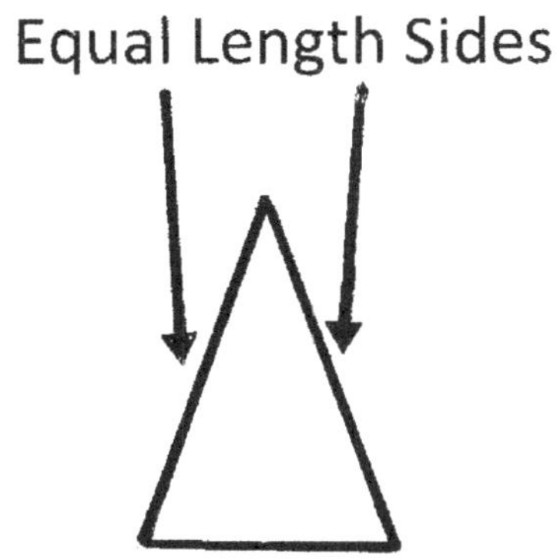

Iterative or iteration – Something being repetitive or repeated

Juxtaposition – To place side by side or close together

Kilo – A metric prefix meaning 1,000 – For example, a kilogram is 1,000 grams

Latitude – The angular distance measured North or South of the earth's equator, where the angle is visualized at the center of the earth of two lines, one going to the equator and the other going to the point whose latitude is being measured – The latitude is 0 degrees at the equator and 90 degrees at either the North or South pole.

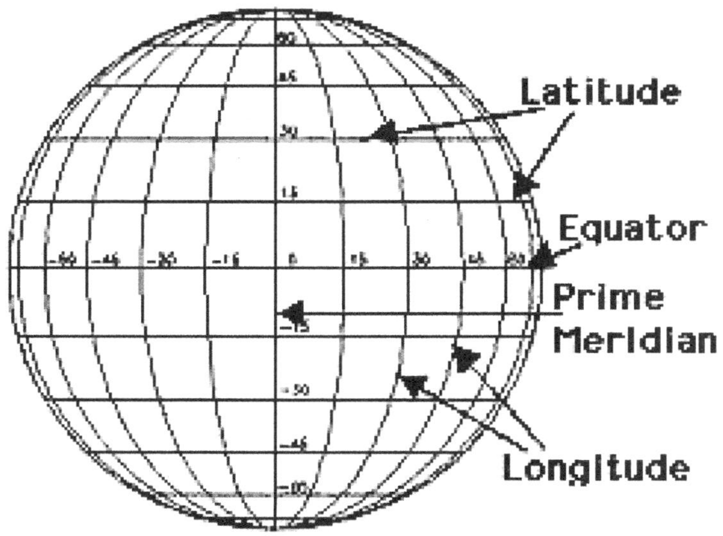

Least Common Denominator – The least or smallest common multiple or factors of the denominators of two or more fractions – For example:

Fraction 1: 3/16 Fraction 2: 1/6

Factors in the denominator of 3/16 are 2 x 2 x 2 x 2 which equals 16

Factors in the denominator of 1/6 are 2 x 3 which equals 6

The factors that are needed to produce either denominator are: 2 x 2 x 2 x 2 x 3 which equals 48

The usefulness of this is to add 3/16 to 1/6 is to convert 3/16 to 9/48 and 1/6 to 8/48

Therefore, 9/48 plus 8/48 equals 17/48.

Legs of a Right Triangle – In a right triangle the two sides that form the right angle vertex or the 90 degree angle vertex

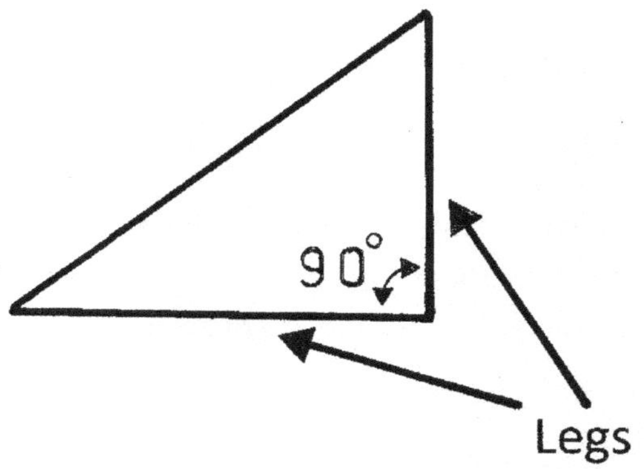

Like terms – Terms (plural) that contain the same variable or variables

Line – Usually understood to be a straight line unless otherwise defined - An infinite set of points forming a straight line extending in two directions to infinity – A Line is drawn sometimes with and sometimes without ending arrows. Either form of drawing is assumed to extend to infinity in both directions unless defined otherwise.

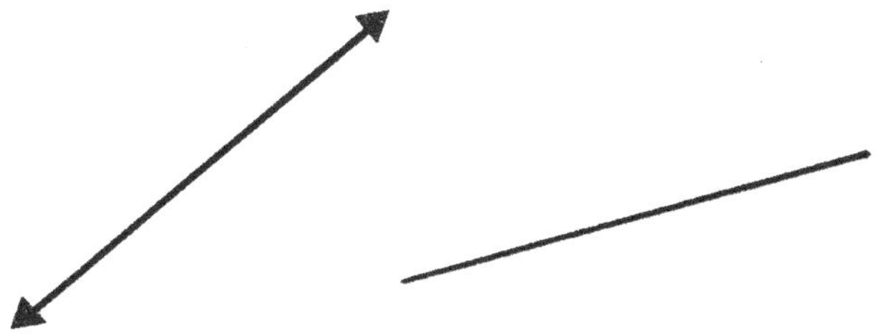

Line of Symmetry – A line that divides a figure or shape into two identical halves, each half being the mirror image of the other

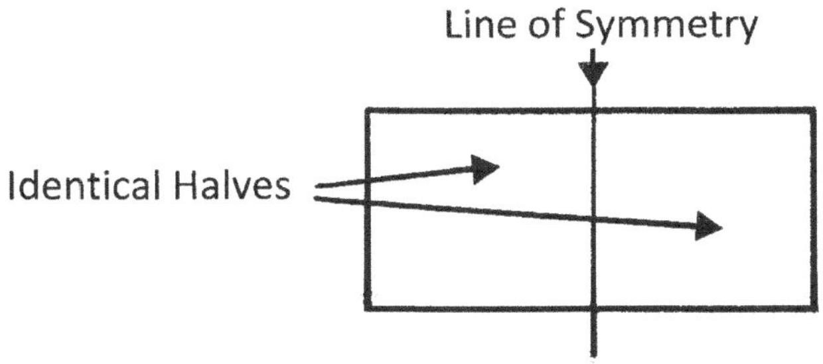

Line segment – A part or section of a line defined by two endpoints – If the Line Segment is symbolized by a letter at each endpoints, such as A and B, then the Line Segment is named as AB with a line over the AB.

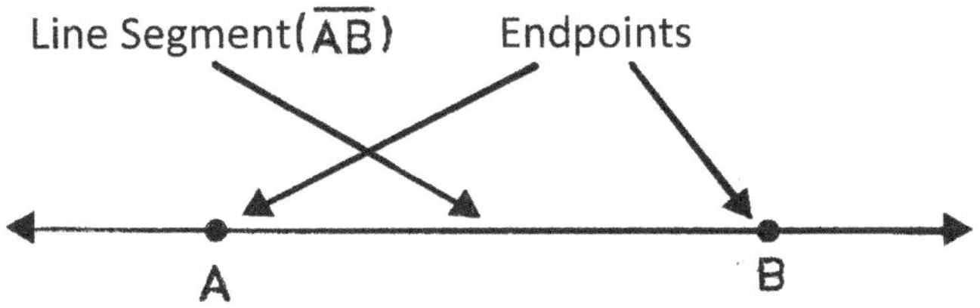

Linear Equation – An equation whose graph in a coordinate system is a straight line – For example, the following is a linear equation in a 2-dimensional coordinate system: $Ax + By = C$

Logarithm – The exponent that expresses the power to which a base is raised to produce another number – For example: for $10^2 = 100$, the logarithm is "2" which is the power to which the base, 10, is raised to produce 100 in this example.

Logic – Formal structure for reasoning

Longitude – The angular distance measured East or West of a line running North and South to the North and South poles, respectively through a point at Greenwich, England, where the angle is visualized at the center of the earth of two lines, one going to the above Greenwich line and the other going to the point whose longitude is being measured – The longitude is 0 degrees at said Greenwich line and 180 degrees at half way around the earth East or West from said Greenwich line. The Prime Meridian (shown below) is the Longitude line that goes through Greenwich, England.

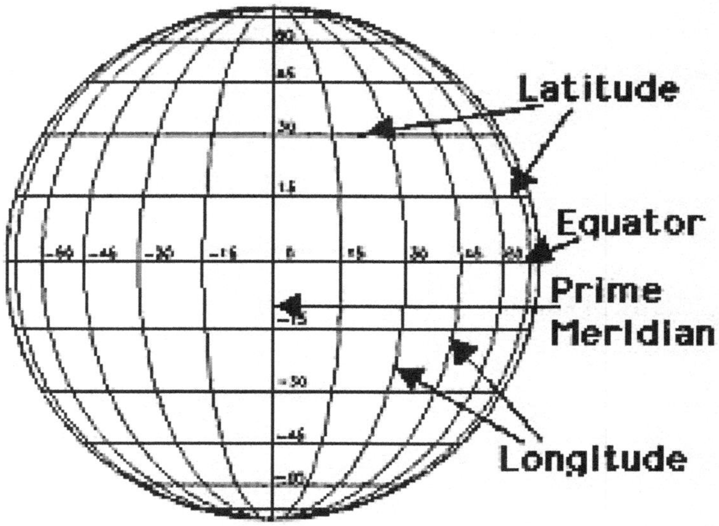

Mantissa – The decimal fraction part of a logarithm value - For example, the logarithm (base 10) of the number 200 equals 2.301 . . . The mantissa of the above logarithm value is .301 . . .

Map – (a) A drawing or other representation usually on a flat surface showing things or interest thereon, or, (b) relating one collection of items or points to another collection of items or points, usually using different representations for each different collection

Mass – The amount of matter in an object

Mathematics – The group of sciences (including arithmetic, geometry, algebra, and calculus) dealing with quantities, magnitudes, forms, and their relationships by the use of numbers, symbols, and operations

Maximum – The largest point value of a function or set of points or values

Mean or Average – The result obtained by adding together a series of numbers and then dividing by the number of numbers in the series – For example, the mean or average of 6 + 5 + 8 + 9 (note 4 numbers in the series) equals 28/4 = 7. "7" is the average or mean of the series: 6 + 5 + 8 + 9.

Median – The middle number found by counting the number of numbers in a set of numbers arranged from least to greatest – frequently not equal to the mean or average – For example, the median of the set of numbers given as "1, 4, 8, 100, 250" is "8" because "8" in this set has two number ("1" and "4" to the left of "8" and two numbers, "100" and "250" to the right of "8".

Mega - A metric prefix meaning a million or 10^6 – For example, a million dollars is 1,000,000 dollars

Metric system – A system of numbers based on intervals of ten, tens, and tenths

Micro – A metric prefix meaning a millionth or 10^{-6} – For example, a micrometer means 1/1,000,000 of a meter, or, 0.000,001 meter.

Midpoint of a line segment – a point that is exactly the same distance from both ends of the line segment and thus half way in position in going from one end to the other end of a line segment

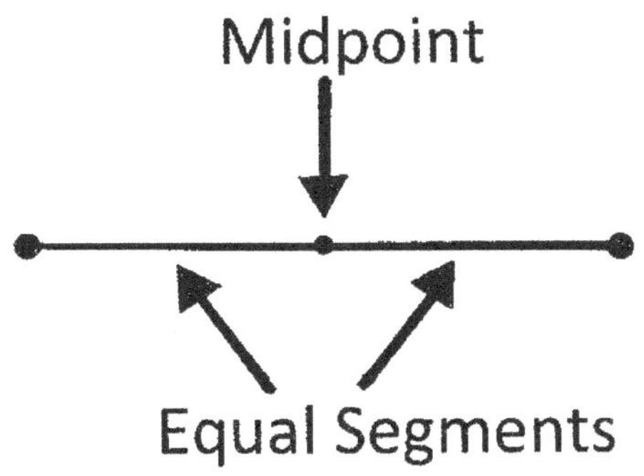

Milli – A metric prefix meaning a thousandth or 10^{-3} – For example, a millimeter means 1/1000 of a meter, or, 0.001 meter.

Minimum – The smallest point value of a function or set of points or values

Minuend – In subtraction, the number that is subtracted from by the Subtrahend to Obtain the Difference – See example:

524	←	Minuend
minus (-) 35	←	Subtrahend
489	←	Difference

Mixed Decimal – A number with an integer and a decimal fraction part: For example: 56.873 ("56" is the integer part and ".873" is the decimal fraction part.)

Mixed Fractional number – A number with an integer number part and a fraction part

For example: 4 3/4

Mode – A number that appears most frequently in a set of numbers

Monomial – (a) In an expression only one term, or, (b) the multiplicative product that may contain numbers and/or variables – Examples of monomials: 3x, 45, 5xyz, or $7xy^3$

Multiple – The multiplicative product of a whole number and another other number, for example, a multiple of 6 is 18, that is, 18 equals 3 times 6

Multiplicand – In multiplication, the number being multiplied by a multiplier – For example, for the multiplication: 2 times 5 equals 10 wherein 5 is the multiplicand, 2 is the multiplier, and 10 is the product

Multiplication – The process of repeated addition – For example 4 multiplied by 5 is 4 + 4 + 4 + 4 + 4 and equals 20

Multiplicative Identity – Multiplying any Number by one or "1" does not change the number – For example, for the number, N: 1 x N = N or N x 1 = N or 1 • N = N or N • 1 = N

Multiplicative Inverse – For any non-zero real number, multiplying the real number by its inverse or reciprocal equals one or "1". For example, for any non-zero number, N: N x 1/N = 1 or 1/N x N = 1 or N • 1/N = 1 or 1/N • N = 1

Multiplier – See above in the definition of Multiplicand.

Mutually exclusive events – Two events that cannot happen at the same time or in the same space

Nano – A metric prefix meaning a billionth or 10^{-9} – For example, a nanometer means 1/1,000,000,000 of a meter, or, 0.000,000,001 meter.

Natural Number – A positive integer used commonly for usual counting – Natural numbers: 1, 2, 3, 4 . . . Zero "0" is not normally understood to be a Natural Number.

Negative Integer – Any Integer which is not a Whole Number – For example: -1, -2, -3, ... Note that Whole Numbers are generally defined a Positive Integers (Natural Numbers) plus Zero.

Negative Number – a number that is less than zero – For example; - 3

Negative sloped line – A line that slants downward from left to right

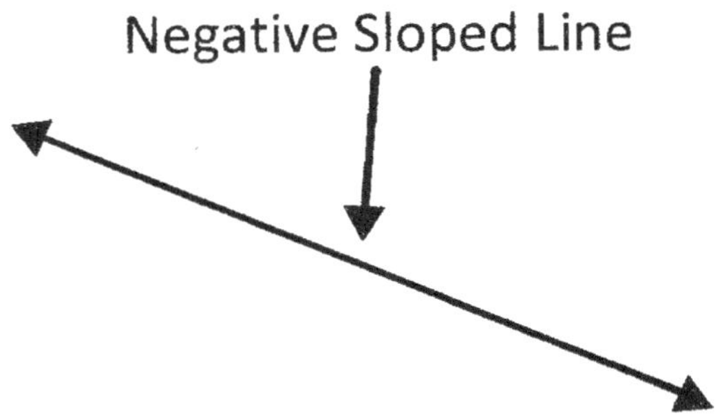

Non-Integer – Any Rational Number that is not an Integer – Examples of Non-Integers are fractions that are not valued when simplified to an integer value.

Non-Repeating Decimal Number – A Non-Repeating Decimal Number does not contain an infinite number of a repeating section made of either digits or a single digit, wherein each repeating digits or digit are identical. A Non-Repeating Decimal Number may be either a Rational or an Irrational Number. A Non-Repeating and Non-Terminating Decimal Number is Irrational.

NonTerminating Decimal Number or **Non-Terminating Decimal Number** – A Decimal Number that contains an infinite number of digits – A Terminating Decimal Number contains a finite; that is, countable; number of digits.

Null – Nothing or empty

Nullify – To reduce to Null or cancel out

Number – A symbol or word showing how many items there are, or, which one is meant in a series of items – Types of numbers: Complex, Real, and Imaginary. There are several types of real numbers; such as rational, irrational, integers, rational non-integers, whole, whole, and natural numbers which are related as drawn below. Specific Number types are defined in detail elsewhere in this Glossary.

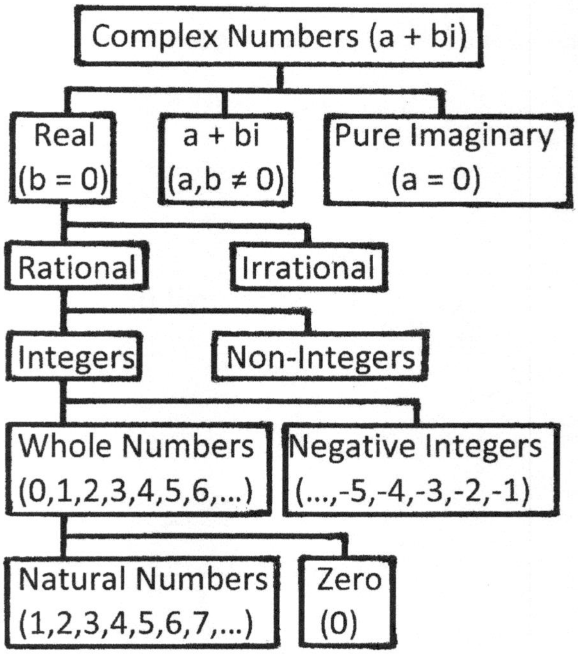

Number line – A line that represents real numbers as points on a line

Here is a drawing of a "Number Line" showing Integers:

. . . -3 -2 -1 0 1 2 3 . . .

Negative Infinity ←——————→ Positive Infinity

Numeral – A number symbol that is not a variable – Examples of numerals: 1, 2, 3, 4

Oblique – A relationship between lines and/or planes that is neither parallel nor perpendicular – See below example of oblique lines:

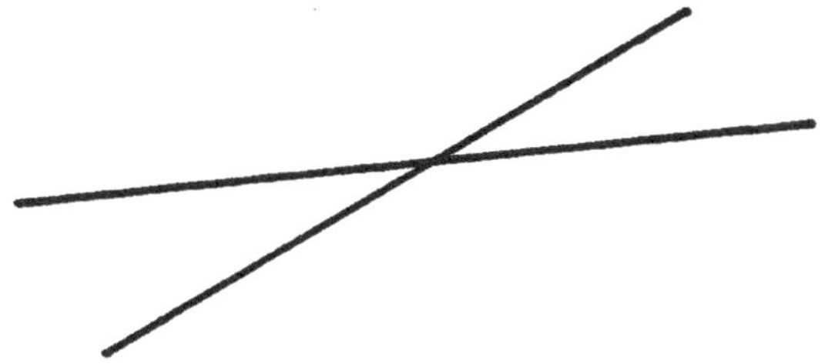

Obtuse Angle – an angle that is greater than 90 degrees (90 degrees is a right angle) and also less than 180 degrees (a straight angle)

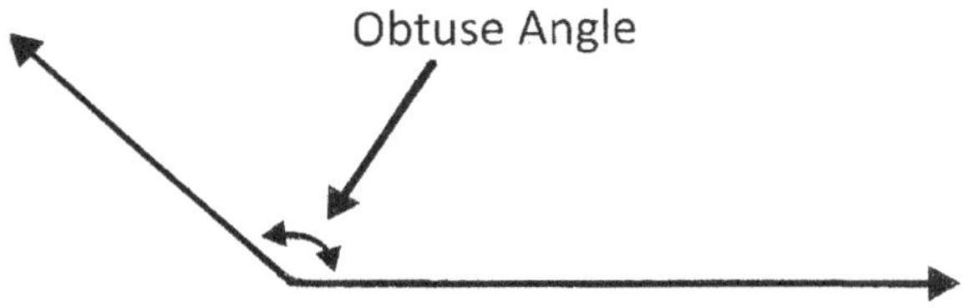

Obtuse Triangle – A triangle whose largest angle therein is greater than 90 degrees

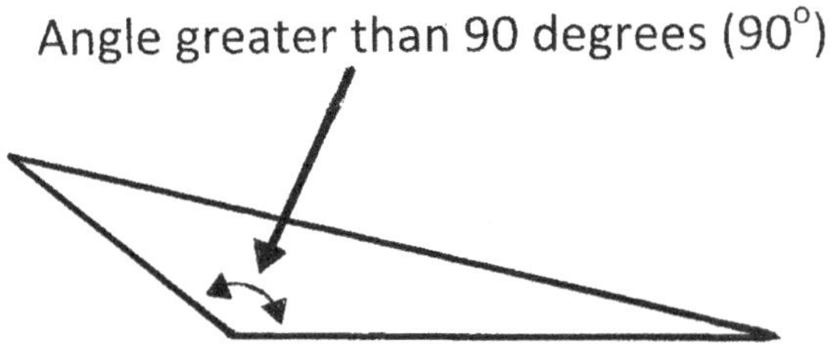

Octagon – A polygon with eight sides

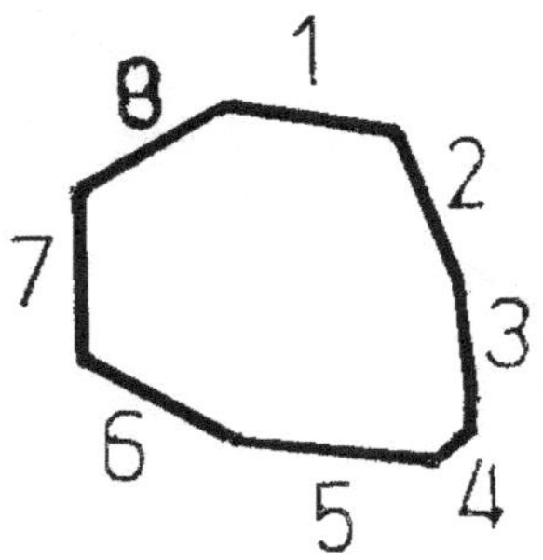

Odd number –(a) A whole number integer that cannot be divided by two to result in another whole number – For example, 1, 3, 5… are odd numbers, or, (b) Sometimes odd numbers are defined as being inclusive of negative integers that cannot be divided by two to result in another negative integer.

Odds – The ratio of favorable to unfavorable outcomes in a set of trials at some desired outcome

Operation – Any process or method involving a change or transformation of an item, term, expression, or quantity into something else

Opposite – (a) Across from one another, or, (b) having a different sign, such as positive versus negative

Opposite Operation – An operation or action of another operation or action – For example, subtraction is the opposite of adding, or, multiplying is the opposite operation of dividing.

Opposite angles of a polygon – Angles in a polygon that do not have any sides in common or the same

Opposite Angles in the below polygon (rectangle)

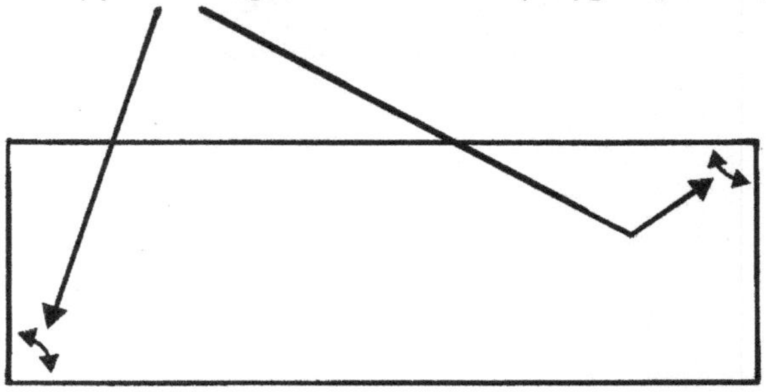

Orbit – A path taken by an object that is usually either circular or elliptical as it travels around another object

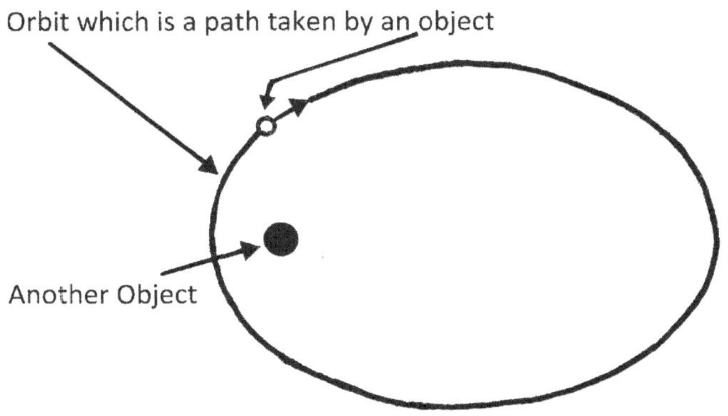

Order of operations – Rules for defining what sequence of mathematical operations are done in order from first to last when evaluating a Mathematical expression. The evaluation order is generally as follows, done in the following order as numbered: (1) Do operations inside groupings; such as parentheses, brackets, etc.; starting with the innermost grouping and going outward from there, using within each grouping the orders hereafter described in order as (2) to (4); (2) Evaluate all exponential expressions; (3) Do multiplications and divisions in order from left to right; (4) Do additions and subtractions in order from left to right. Also, any operations in a fraction are separately done above and below the fraction bar or line before evaluating the fraction operation, if possible. For example, to evaluate the expression: $2 + 5 \bullet 3$; the $5 \bullet 3$ is done first to get 15, and then 2 is added to 15 to get the result which is 17.

Ordinate – the vertical (in the y-axis direction) Cartesian coordinate on a plane, measured from the x-axis along a line parallel with the y-axis to a point. It may be a positive number, if measured in the positive direction from the x-axis or it may be a negative number, if measured in the negative direction from the x-axis. In the below diagram the point coordinate numbers (1,2) is one unit to the right from the y-axis and therefore the abscissa of the point is "+1". The coordinate number "2" is 2 units up from the x-axis and is called the ordinate, valued at "+2" for the below point. An ordinate is also the vertical (in the y-axis direction) coordinate used in a 3-dimensional (x,y,z) coordinate system.

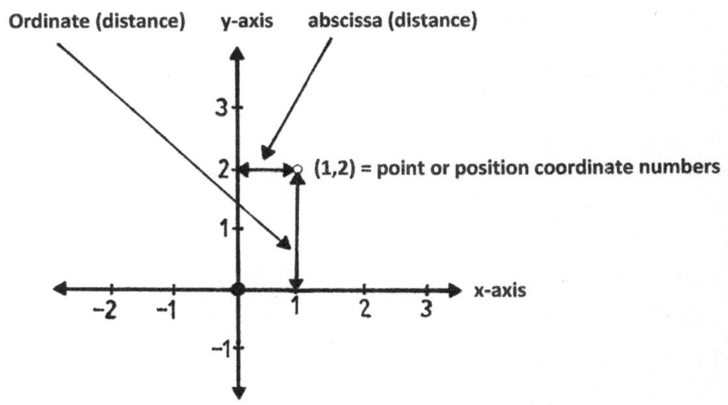

** Note: When defining the location or placement of a point, such as "o" above, on a coordinate system such as used above, the abscissa is always written before the ordinate with a comma between them and within parentheses, such as "(1,2)" above.

Origin of a coordinate system – The point where the axes (x-axis and y-axis above) that define the coordinate system positions all converge together at a point, where the position coordinates generally are zero (0) in value – See the x-axis/y-axis intersection point "•" above where the x-axis and y-axis lines cross.

Orthogonal – (a) Perpendicular, or, (b) relating to right or 90 degree angles

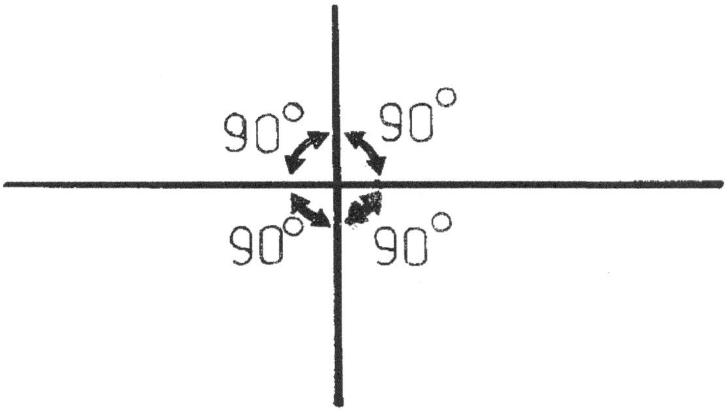

Outcome – One of many possible events in a situation involving probability

Outlier – A data point in a set of data that is much larger, smaller, or distant compared to most of the other data points

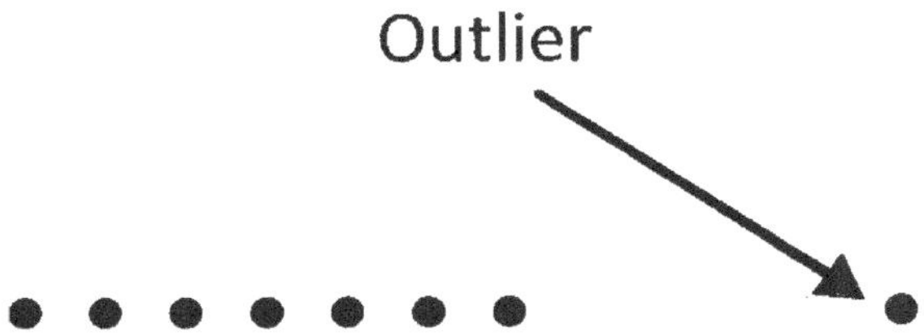

Overestimate – An estimate that is greater than the exact or correct result or answer

Parabola – (1) A planar figure or shape formed by points that are all the same distance from a fixed point and a fixed line (perpendicular distance), or, (2) a conic figure formed by the intersection of a plane with a cone wherein the plane is parallel with one side of the cone

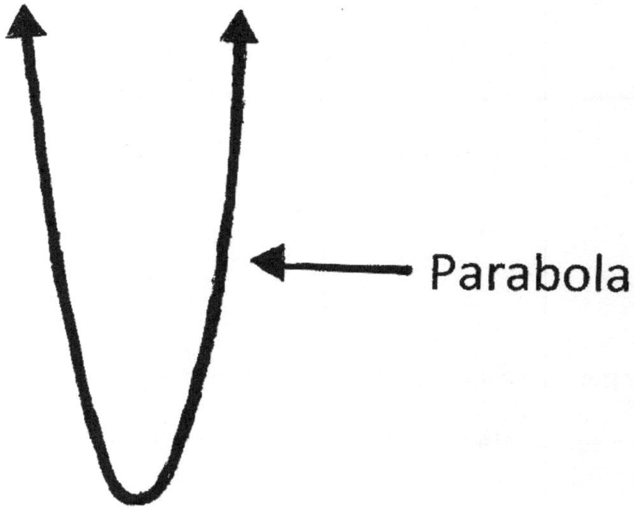

Parallel – A set of points that is always the same perpendicular distance from another set of points – For example, these line segments are parallel: ≡ or ‖ or | |.

Parallelepiped – A polyhedron (3-dimensional object) with six faces in which opposite faces are identical parallelograms, rectangles, or squares

Parallelogram – A (2-dimensional) quadrilateral where opposite sides have the same length and are parallel

Pentagon – A polygon that has five sides

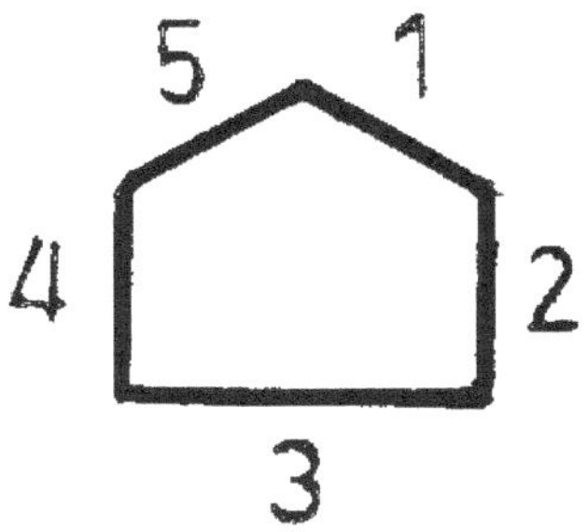

Percent or Percentage – A number expressed as a ratio compared to 100, or, divided by 100 – For example, 40 is 40% of 100, or, expressed in fractional terms 40/100 (a representation of 40%) = 0.40 and therefore 40% or 100 equals 0.40 times 100 which equals 40. Also, for example, 40% of 800 equals 0.40 times 800 which equals 320

Perimeter – The distance around a figure or shape – For example, adding the length of all of the sides of a figure or shape

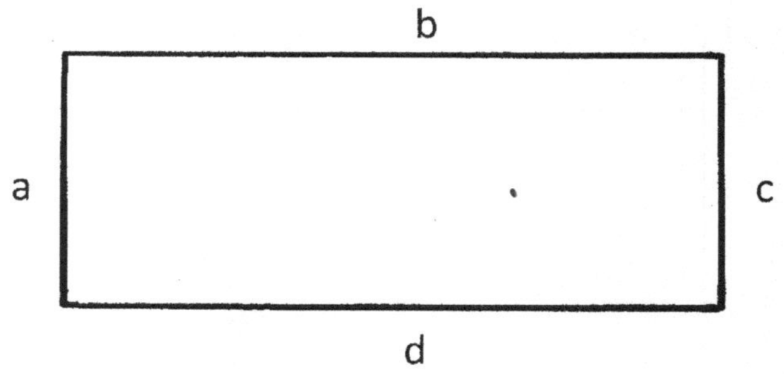

The perimeter of the above rectangle equals a + b + c + d

Period – (a) the length of time for each cycle of a repeating event, or, (b) a general length of time or distance, or, (c) the symbol that separates the whole number part from the decimal number part of a decimal number with both of those parts, that is, "."

Periodic Functions – Functions that form a periodic result or graph

Sine Wave Graph

Sawtooth Graph

Square Wave Graph

Permutations – Different orders of a set of data or series of events

Two permutations of a set of numbers are:

First permutation: 5 6 8 9

Second permutation: 8 6 5 9

Perpendicular – Two line or items that forming a right or 90 degree angle at a corner between them

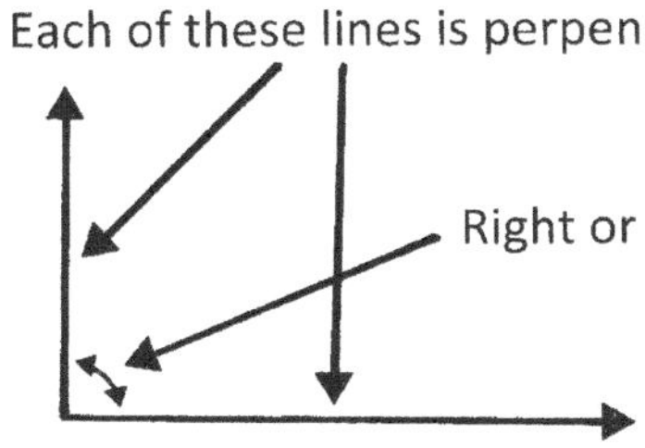

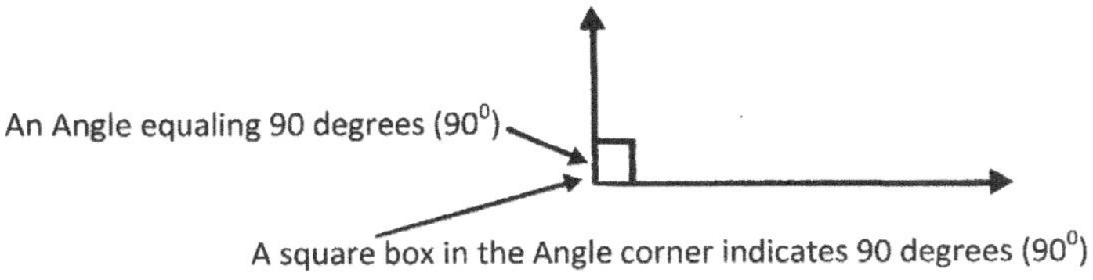

Perpendicular bisector – A line that divides a line segment exactly in half and is perpendicular to that line segment

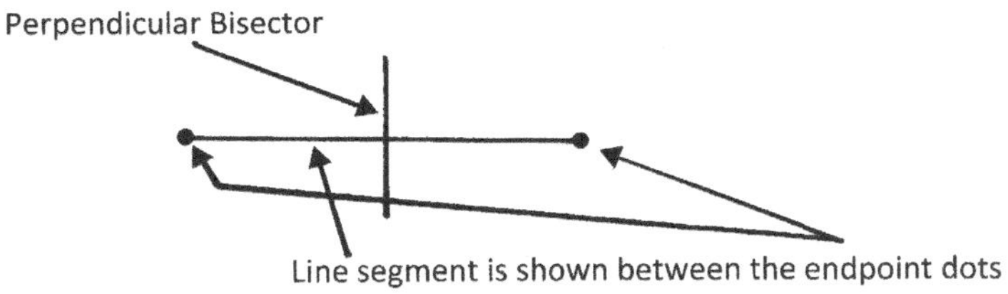

Note: All four of the angles at the corners are right or 90 degrees (90⁰)

Perpendicular distance – The length of a line segment that goes from a point to a line along a line that passes to the point and is perpendicular to the line segment

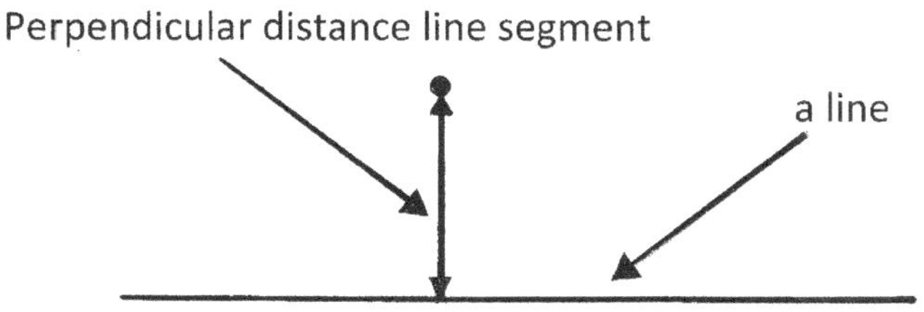

<u>Pi</u> – symbolized by the Greek letter pi (π) and is the ratio of the circumference of a circle to its diameter, equal to approximately 3.14159 . . .

<u>Pico</u> – A metric prefix meaning a trillionth or 10^{-12} – For example, a picometer means 1/1,000,000,000,000 of a meter, or, 0.000,000,000,001 meter.

<u>Place value</u> – The numeric value of a number symbol defined by its "place" in the number. For example, in the whole number 457 the place value of the 5 is 50 and the place value of the 4 is 400. The rightmost number place only indicates its number value itself and no added place value. Each further added place to the left adds a multiplier of 10 to the place value of a number. A zero "0" is sometimes in a number and it indicates that there is no numeric value for that "place" but that the "0" is needed to give numbers further to the left their proper place value. For example, the value of 204 is "4" plus "200". The zero "0" in the middle of "204" only keeps the place value of the left-hand 2 at "200". The zero itself has no numeric value other than its own zero "0" value. Additionally, if a number is written as a decimal number than the "decimal point", shown as ".", is written to the right of the rightmost least place value number. Therefore, "204" has the same value as "204.". "204." is written as a "Decimal Number" and "204 is written as a Whole Number, Integer, or Natural Number.

<u>Planar figure</u> – A geometric shape or figure wherein all of its parts are contained in a single flat plane – For example, a circle, square, or polygon are generally planar figures

<u>Plane</u> – any two intersecting lines in space determine or define the location of a plane – A plane separates a three dimensional space into two parts, called half spaces – note that a plane generally extends out to infinity in all directions and is a flat surface

<u>Point</u> – An exact position in a space – For example: •

<u>Polygon</u> – A closed figure or shape that is formed from line segments that meet only at their endpoints

Triangle (3 Sides)

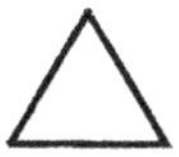

Pentagon(5 sides)

Hexagon (6 sides)

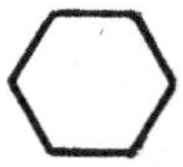

Quadrilateral or Tetragon or Quadrangle(4 sides)

Polyhedron – A three-dimensional figure or shape wherein all of its sides are polygons – See below example which is a cube polyhedron.

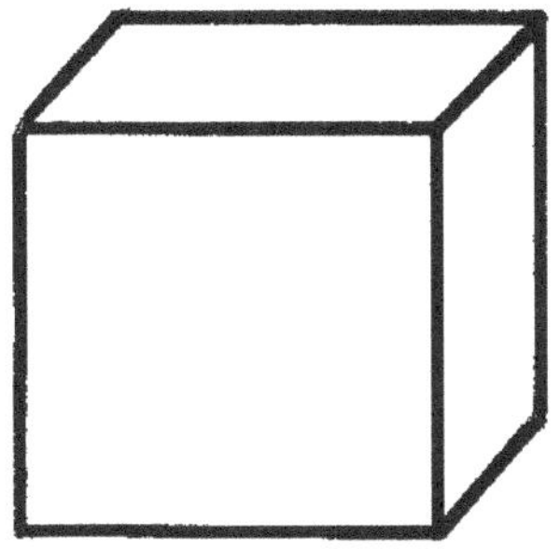

Polynomial – An expression containing a plurality of monomials – For example: 3x + 4y + 5z

Position – (a)The location of a point, or, (b) coordinates that define the location of a point

Positive Number – A number that is greater than zero, "0"

Positive sloped line – A line that slants upward from left to right

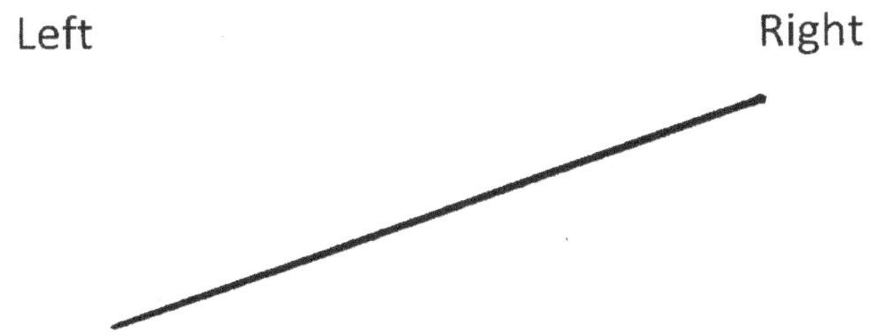

Power – An exponent – For example, the exponent in the number 4^3 is "3".

Precision – The accuracy or exactness of a measurement – minimizing errors

Premise - (a) A basis for further investigation; or; (b) An unproved theory; or; (c) A tentatively or temporarily accepted proposition or proposal to explain certain facts; or; (d) A hypothesis

Prime Factorization – Expressing a number as a multiplicative product of prime numbers: For example, the prime factorization of the number 14 is 2 • 7. "2" and "7" are prime numbers.

Prime Number – A number that has exactly only two positive factors, whole numbers, which are itself and one, or, a number that can only be divided by itself and one (1) to result in a whole number – Note, the number "1" is not a prime number because it cannot be divided by two whole numbers to result in a whole number because only itself "1", one whole number is a divisor of "1" to result in a whole number.

Probability – The ratio of the number of times that something will probably occur to the total number of possible occurrences

Product – The result of a multiplication operation - In multiplication - For example in the expression: 2 times 5 equals 10; 5 is the multiplicand, 2 is the multiplier, and 10 is the product

Proper Fraction – A fraction whose numerator is an integer that is smaller than the integer which is the denominator – For example, 4/9 is a proper fraction

Proportion – An equation showing that two ratios are equivalent – For example, 4/9 is proportional to the ratio 8/18 because both the numerator and denominator of the second ratio or fraction are both doubled so that for every 4 in the numerator there is a corresponding 9 in the denominator

Proportional – Having equivalent ratios – See 4/9 being equivalent as a ratio to 8/18 above.

Proposition – (a) An expression or sentence that is either true or false, but not both; or; (b) A building block for a theory; or; (c) An expression or statement that can be evaluated whether it is true or false

Protractor – An instrument in the form of a graduated semi-circle, used for plotting and measuring angles – See the below Protractor with degrees indicated with numbers 10 degrees to 170 degrees. – Protractors also are made in other forms, such as full circular types, for example. Some protractors are graduated in degrees, but also can be graduated in radians (Radians are another type of measurements of an angle.)

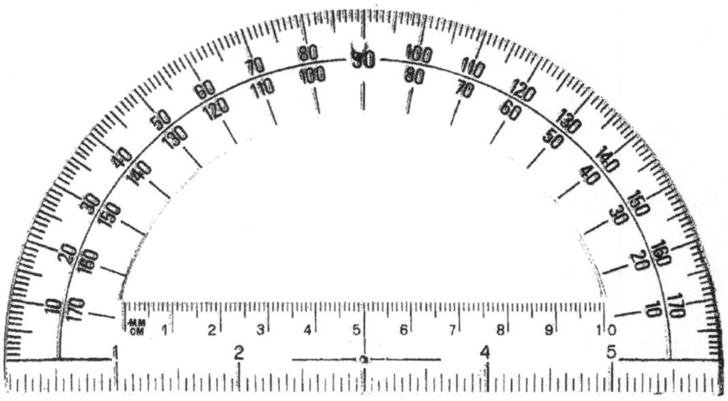

Pyramid – A polyhedron whose base is a polygon and whose faces are triangles that share the same vertex

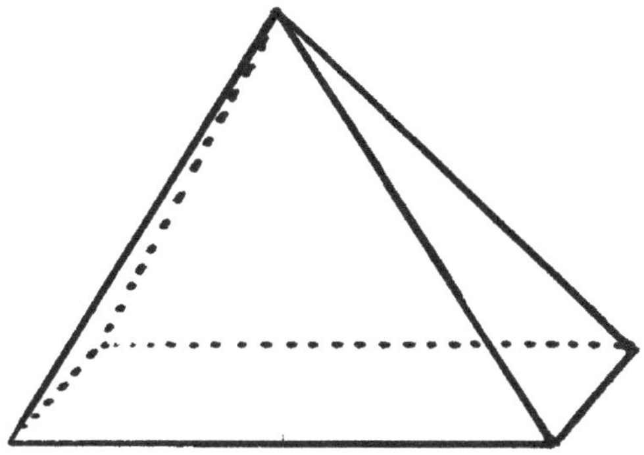

Pythagorean Theorem – The sum of the squares of the lengths of the legs (smaller sides) of a right triangle equals the square of the length of the hypotenuse – (one leg length)^2 + (other leg length)^2 equals (length of the hypotenuse)^2

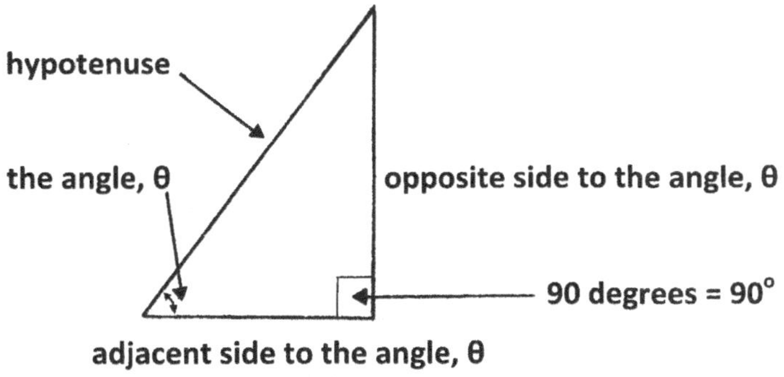

Pythagorean Triples – Sets of three positive integers that fit the Pythagorean Theorem lengths – For example; 3, 4, and 5 is a Pythagorean Triple because $3^2 + 4^2 = 5^2$. The leg lengths are "3" and "4" and the hypotenuse length is "5" for this Pythagorean Triple.

Quadrants – The four sections of a Cartesian coordinate (two-dimensional grid system) that are separated by the coordinate axes

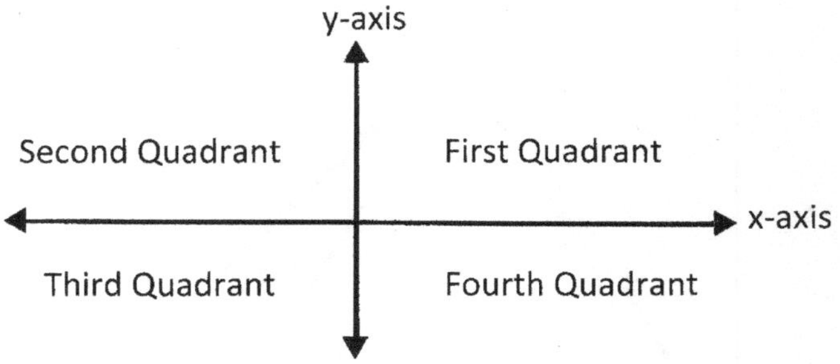

Quadratic Equation – An equation containing terms wherein the highest power or exponent therein is a square or power or exponent of two for any variable – For example: $Ax^2 + Bx + C = 0$

Quadrilateral – A closed 2-dimensional figure (that is, in a plane) with 4 sides that each are straight line segments

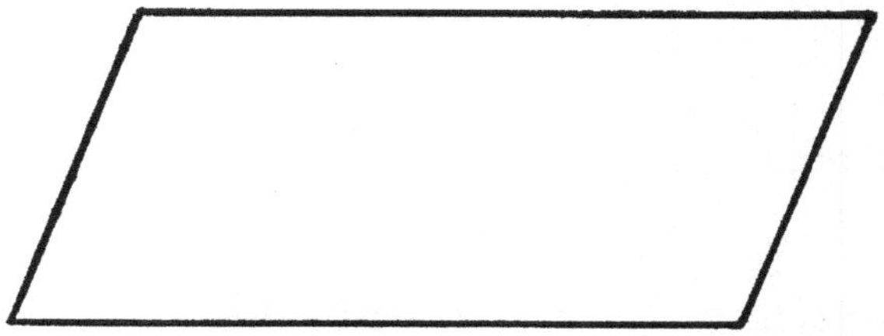

Quantity – An amount of something

Quarter turn – One-fourth of a full revolution, that is turning 90 degrees (90^0) around a point, center, or line

Quartiles – Quartiles divide a set of data into about four equal sets of data

Quotient – The result obtained when one number is divided by another number

Radian – The angle formed at the center of a circle by two radii lines cutting or defining an arc on the circle equal in length, along the arc, to the radius of the circle – 1 radian equals 57.2957795 . . . degrees and the total of all degrees at the center of a complete circle is 360 degrees – those 360 degrees also equals 2π radians.

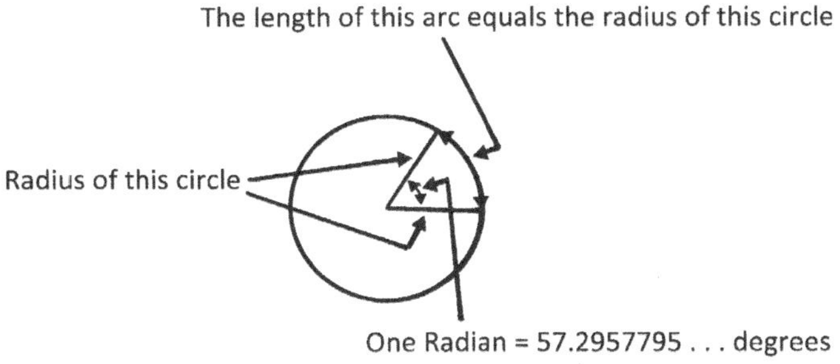

Radical – The symbol "√" indicating a root of a number or expression which is written under the radical symbol – without a superscript, the square root is understood – with a superscript the root number is superscripted – For example $^3\sqrt{x}$ indicates that what is meant is the cube root of "x".

Radicand – The number or expression under the radical symbol of which the root is desired – In the above example: $^3\sqrt{x}$ the "x" is the radicand.

Radii – straight lines connecting the center of a circle or sphere with the circle or sphere – plural of radius

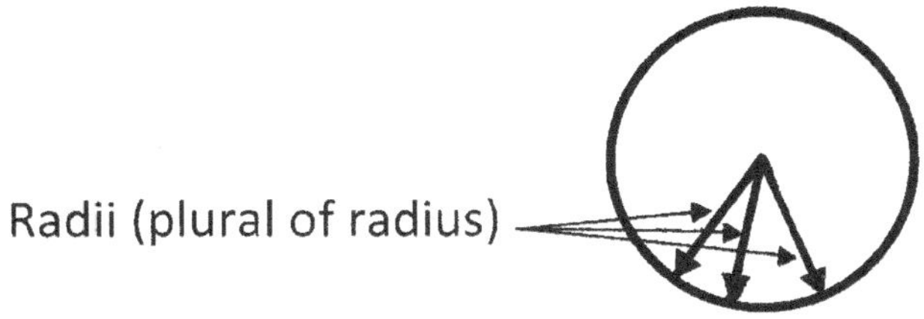

Radius of a Circle – The radius of a circle is the fixed distance of all points on a circle wherein the fixed distance is the same distance from a fixed point, called the center of the circle – See above drawing. The diameter of a circle is two radii in length.

Radius of a Sphere – The fixed distance of all points on a sphere wherein the fixed distance is the same distance from a fixed point, called the center of the sphere – The diameter of a sphere is two radii in length

Random – All outcomes of an event having an equal chance or likelihood of happening

Range – (a) The difference between the smallest and largest in a set of numbers, or, (b) The possible values that a variable can have in a particular situation, or, (c) the possible values that a function can have

Rate – A product of dividing one quantity (dividend) by another quantity (divisor) wherein the divisor is usually a time interval – For example, miles per hour is a rate.

Ratio – Comparing two quantities or numbers by either division or by stating the two quantities or numbers either as a fraction or separated by a colon ":" – Examples of ratios are: 3/8, 3:8, and 12 Ford cars sold per 100 total cars sold in a day gives the ratio of 12:100

Rational number – A real number expressible as an integer or as the quotient, or ratio, of two integers; wherein the denominator is non-zero – A fraction made of two integers is rational. A repeating decimal number is rational. A decimal number that terminates is rational.

Ray – A directed line in one direction – See example below. The name of an array commonly is designated with a part arrow or arrow above the name.

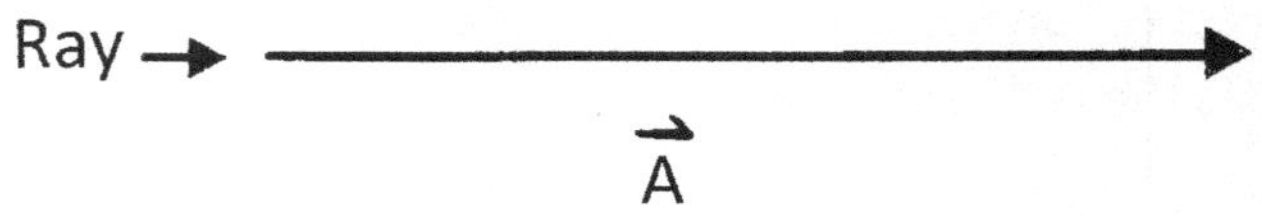

Real Number – Any number that is either rational or irrational – A real number is generally understood to not contain an imaginary part.

Rectangle – A quadrilateral which is a Parallelogram with all corners being right (90 degree) angles

Rectangular Coordinate System - A three dimensional coordinate system wherein the coordinates are defined by three perpendicular axes, sometimes called a Cartesian Coordinate System. Points in space are defined in such a coordinate system by three numbers, usually x, y, and z values – See above definition of a Cartesian Coordinate System

Rectangular Coordinates – Coordinates that define a point's position in a Rectangular Coordinate System or Cartesian Coordinate System - See above.

Recurring Decimal or Repeating Decimal – A decimal made of repeating numbers – For example, a recurring or repeating decimal is 0.181818181818 . . . because the number pair "18" repeats – A Recurring Decimal or Repeating Decimal is a type of Number called Rational.

Reducing a fraction – Simplifying a fraction into its simplest form, usually by cancelling out any factors that are present both in the numerator and denominator

For example, 3/6 can be written as (3 x 1) divided by (3 x 2).

$$\frac{3 \times 1}{3 \times 2} = \frac{\cancel{3} \times 1}{\cancel{3} \times 2} = \frac{1}{2}$$

The numerator and the denominator both contain the number "3" as a factor. These "3"s can be removed or canceled from both the numerator and denominator without changing the value of the fraction. After removing or cancelling the 3s the fraction left is ½. No further simplification is possible because there are no factors that can be removed or cancelled from the numerator and denominator without changing the fraction further. The fraction ½ could be written as 1 times [1 divided by (1 times 2)]. Cancelling or removing the 1s from the numerator and denominator leaves still ½ and so this fraction is unchanged and cannot be further simplified.

Reflection – Making a mirror image of something such as a figure or shape

Reflex angle – An angle that is more than 180 degrees (180^0)

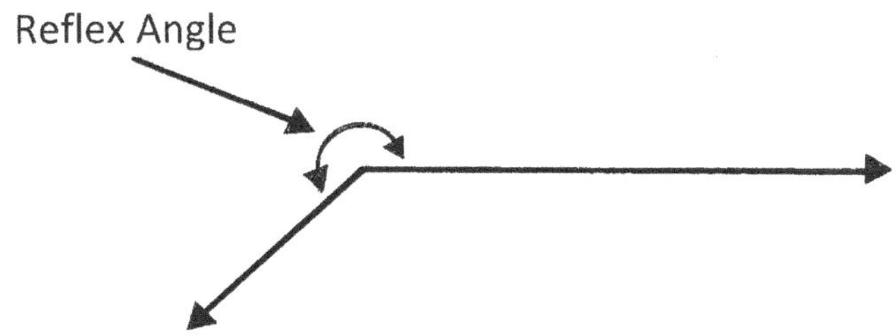

Reflexive Property – A number equals itself, that is, a = a

Regular polygon –A polygon with all sides the same length and all angles the same – For example, a square is an example of a regular polygon

Regular polyhedron – A solid figure with all sides the being made of the same regular polygons – For example, a cube is an example of a regular polyhedron

Relation – A correspondence between two sets, such as a set of ordered pairs or coordinates – The set of all first coordinates is the domain of the relation, and the set of all second coordinates is the range of the relation.

Remainder – In whole number division, when the division has gone as far as it can go without using decimals, any remaining number is the remainder

Repeating Decimal Number – A Repeating Decimal Number contains an infinite number of a repeating section made of either digits or a single digit, wherein each repeating digits or digit are identical. A Repeating Decimal Number is a Rational Number. See the below example of a Repeating Decimal Number which equals 1/3.

$$0.\overline{3} = 0.3333333333333\ldots$$

Reverse – Change direction to the opposite of what was the original direction

Revolution – One complete turn of 360 degrees (360^0) around a point

Rhombus – A polygon with all sides being the same length – For example a square is a rhombus as well as the following:

Right Angle – An angle that is a perfect corner at 90 degrees (90^0)

Right Triangle – A Triangle containing one angle which is 90 degrees (90^0)

Rise – The increase in a y direction value between two points on a graph of a line def ned by y versus x

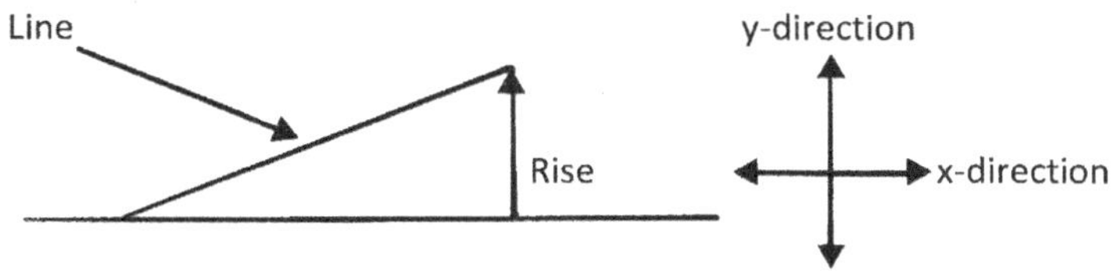

Root – One of the roots such as square root, cube root, etc. that when multiplied together the number of times of the root number gives the number of which the root has been determined - For example, if 2 is multiplied together as 2 times 2 it equals 4. Therefore the square, or second root, of 4 is 2.

Rotate or Rotation – Turning a geometric figure around a point or line

Round or Rounding of a number – To drop or zero out numerical digits to the right in a number – Rounding off or dropping numbers that are 5 or greater is done by adding 1 to the digit just left of the dropped portion of a number. Rounding off or dropping numbers that are less than 5 is done by simply dropping or zeroing out those dropped numbers. For example, rounding 674 to the nearest tens gives 670 as the result. Rounding 676 to the nearest tens gives 680 as the result. A decimal number is rounded similarly. For example, rounding 45.854 to the nearest tenths gives 45.9. Rounding 45.844 to the nearest tenths gives 45.8.

Run – The difference in x direction values between two points in a graph of a line. For example, for the two points in x-y coordinates of (x,y) being (3,8) and (9,6), the run is 9 minus 3 which is 6.

Sample – A selection of items from a group or population that is chosen to be representative of the whole group or population

Scalar – A number or quantity that has a magnitude or numerical value but no direction associated with it – For example, 3 Apples is a scalar

Scalene triangle – A triangle with no two sides being of the same length

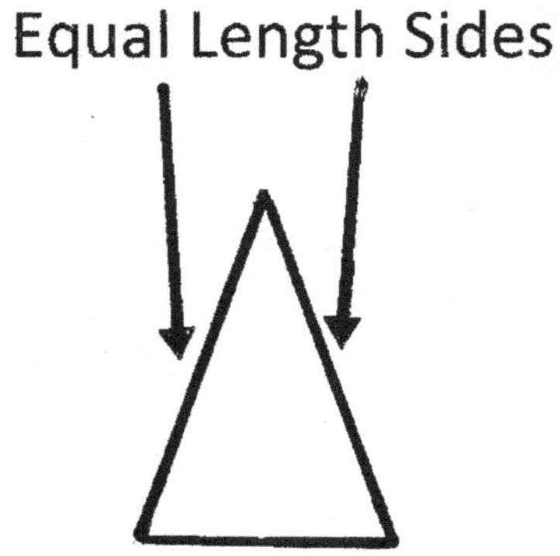

Scaling – Changing all of the lengths of an item or object by the same multiplicative factor – For example, if a length of 20 feet is scaled down by ½ then the scaled length is ½ times 20 which equals 10.

Scientific Notation – Writing a number as the product of a decimal number greater than or equal to 1 and less than ten times an power of 10 exponential number – For example, 467 written in scientific notation is 4.67×10^2.

Sector – A portion of a circle bounded by two radii and the arc formed between where the two radii meet the circle

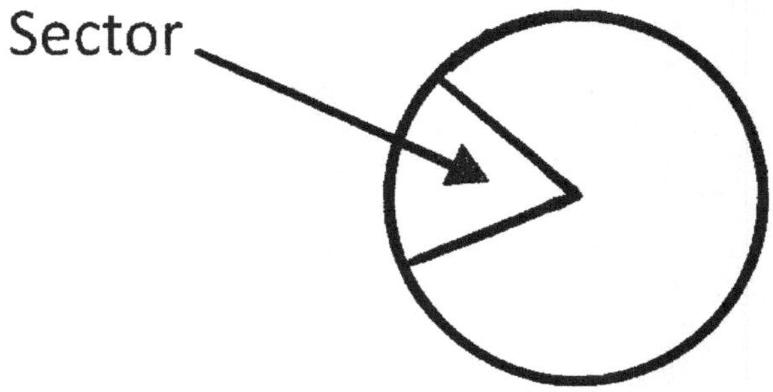

Segment of a circle – A minor segment of a circle is a section of a circle area bounded by a chord and the arc formed between where the chord meets the circle and which is less than half of the interior area of the circle. A major segment of a circle is a section of a circle area bounded by a chord and the arc formed between where the chord meets the circle and which is more than half of the interior area of the circle.

Semicircle – Exactly half of a circle

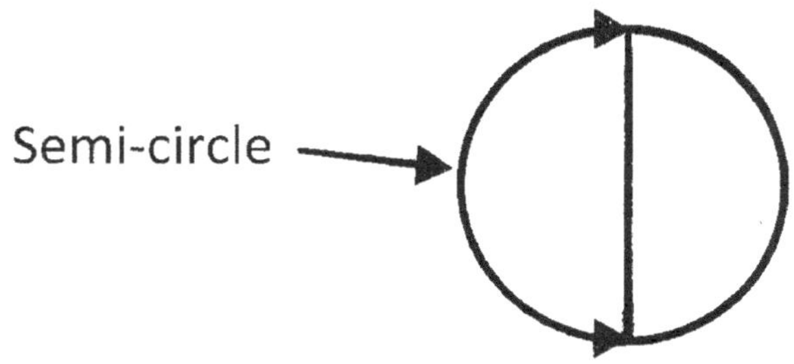

Septagon or Heptagon – A planar figure with seven sides and seven angles

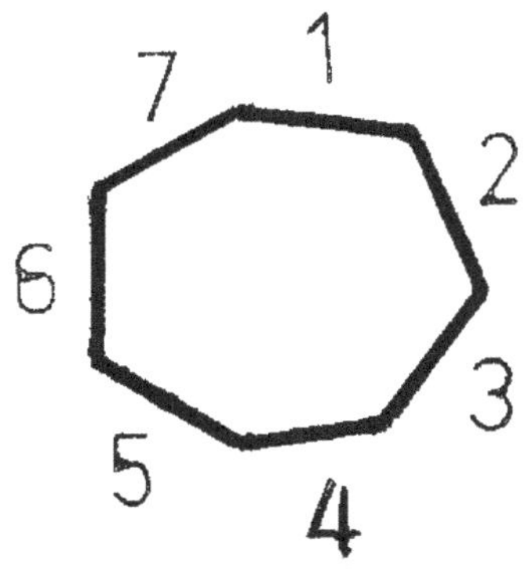

Sequence – Numbers arranged in a special order or pattern

Series – The sum or product of a sequence of component numbers or symbols for numbers. A series may be either of finite or defined length or of infinite length

Set – A designated collection of objects, items, or terms – The designation is a listing of the members of the set of by defining some property possessed by each member of the collection

Signed Number – A number that is written to state that it is either a positive or negative number in value – For example, +4 and -8 are both signed numbers

Significant Digit – A digit in a measured number that can be used with confidence that it is accurate – The most significant digit is the left-most digit that is non-zero. The least significant digit is the right-most digit that can be used with confidence.

Similar figures – Figures or shapes that may be of different sizes but have corresponding sides that are proportional in length with the same proportional differences between corresponding sides – For example, a square with sides all 2 feet in length is similar to a twice larger square which has all sides that are 4 feet in length.

Simple Fraction or Common Fraction – A fraction whose numerator and denominator are both whole numbers – For example, these are simple or common fractions: 5/7, 6/13, and 45/9

Simplest Form Fraction – A fraction whose numerator and denominator do not have any factors that are the same, both in the numerator and denominator, except for one (1)

Slope – The slope of a tangent line at a designated point to a given curve or surface. For example, for a line or curve shown in Cartesian coordinates on a two dimensional (x,y) graph, the slope at a designated point is the ratio of the change in y to the change in x of a tangent line touching or drawn to the line or curve at the designated point

The slope of this Line is Δy/Δx

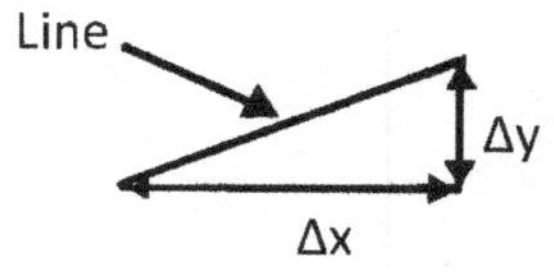

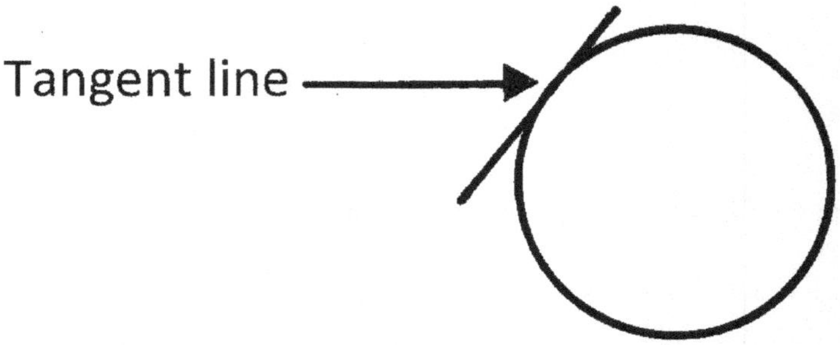

Slope of a function – The slope of a tangent line to the values of a function at a defined point, usually shown on a graph or drawing of the function – See above tangent line which has a slope as defined above.

Solution of an equation – Any value of a variable or variables in the equation that makes the equation properly equating both sides of the equation when the variable is substituted by the solution numerical value – For example, for equation: $2 \bullet x = 6$ the solution is that $x = 3$; that is, substituting 3 in the equation for x changes the equation to: $2 \bullet 3 = 6$

Solution of a System of equations – Values for variables in the system of equations that solve all equations in the system

Space – A defined region in which something may exist and may be defined as to its location. Such a region may be one dimensional as for the number line, two dimensional as in a plane, three dimensional as in a region defined by x, y, and z coordinate values compared to fixed axes. A set of axes, however, themselves may be movable. Another space may include both position and time and is sometimes called 4-space where things are defined by x, y, x, and t coordinate values.

Sphere – a figure or locus of all points which are at the same fixed distance from a point, called the center – The fixed distance is called the radius of the sphere – an equation for a sphere can be as follows: $(x - h)^2 + (y - k)^2 + (z - l)^2 = r^2$; wherein (x, y, z) are Rectangular or Cartesian coordinates of the points on the sphere surface, (h, k, l) are the Rectangular or Cartesian coordinates of the center of the sphere, and r is the radius of the sphere; another form for the equation of a sphere is alternatively given as $x^2 + y^2 + z^2 + Dx = Ey + Fz + G = 0$, however, D, E, F, and G are not so easily identified regarding the sphere's characteristics as the (x, y, z); (h, k, l); and r are in the previous equation for a sphere

Square – A polygon or plane figure defined by four equal length sides and four corners, each a right angle or 90 degree angle

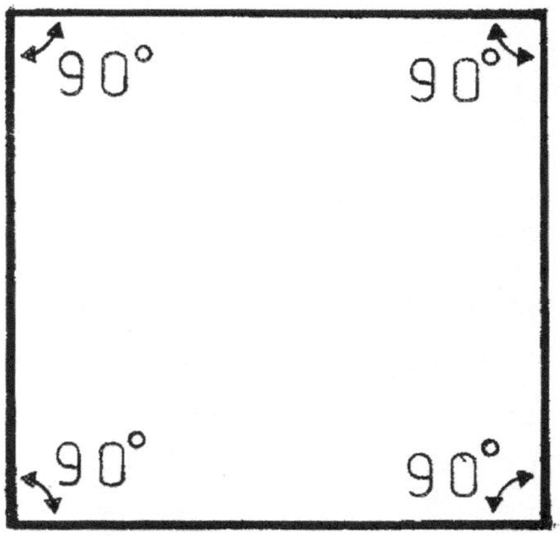

Square of a number – a number that is squared or raised via the exponent "2" to what is termed the second power. For example, 2 squared is designated 2^2 which is 2 times 2 which equals 4

Square root – A number, called a square root, which when squared, or raised via the exponent "2", results in the number for which the square root is desired. For example, the square root of 4 is designated $\sqrt{4}$ and has the value "2". That is, the square root which is "2", when squared, means 2 times 2. This square root of 4 being "2", when multiplied once by itself, or squared (2 times 2), results in the number 4. As an equality expression, this is expressed as $\sqrt{4} = 2$. This could be expressed as $^2\sqrt{4} = 2$, but if the superscript ^{2} in $^2\sqrt{4}$ is not written it is generally assumed to be ^{2} because the most common root of a number is the square root. Higher numbered roots also sometimes, infrequently, are desired, such as the cube root of a number, such as X, written as: $^3\sqrt{X}$, or the 4th root of a number, such as X, written as: $^4\sqrt{X}$. The Nth root of a number, such as X, would be written as $^N\sqrt{X}$, or, sometimes written equivalently using an exponential type of expression written as $X^{1/N}$.

Statement – An assertion that is either true or false

Straight Angle – An angle which is 180 degrees (180^0), or, where two straight lines meet at a common point but both lay along the same combined straight line, or, a line without curves therein

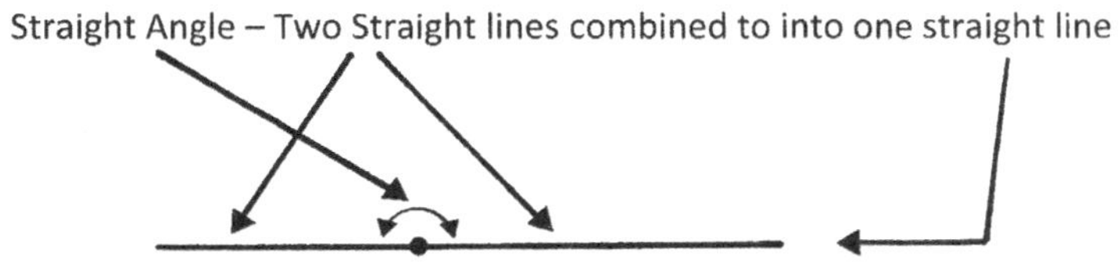

Subtraction – The operation that results in the difference between two numbers or quantities, the result also called the difference – Difference = Minuend minus Subtrahend

Subtrahend – The number being subtracted from another number, called the minuend

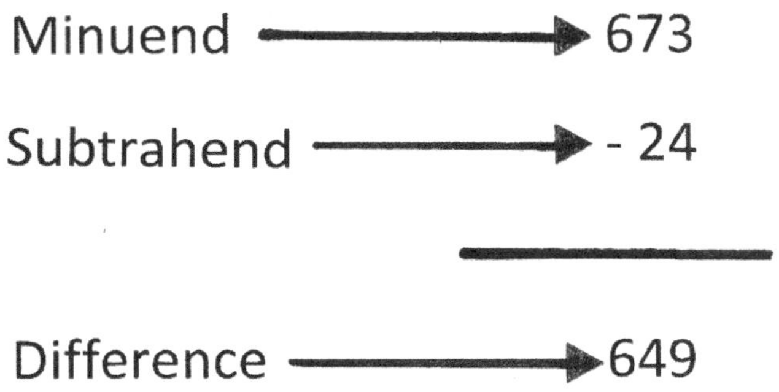

Sum – The number that results from adding two or more numbers

Summation – The operation of adding two or more numbers

Supplemental angles – Two angles that, when added together, result in the summation of them being 180 degrees (180^0)

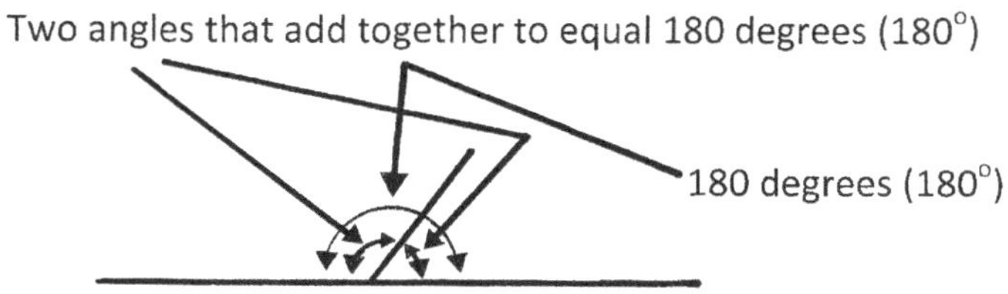

Surface area – The area of a surface, generally represented in square units – For example, the surface area of a 4 foot x 4 foot square is 16 square feet = 16 ft^2.

Symbol – A figure or shape that represents something or another thing

Symmetric or Symmetrical – (a) Terms or elements that can be interchanged without changing the value of either term or element, or, (b) A thing that can be divided into similar halves by a line or plane passing through the middle of the thing

Symmetric Property – If a = b, then b = a.

Symmetry – A term, element, or thing that is symmetric or symmetrical

Tangent – (a) Something that touches something else at exactly one point, or, (b) A Geometry term related to a right triangle wherein the Tangent of an angle therein is the ratio of the opposite to the adjacent side to the angle. See the above Tan θ in the symbol list.

Term – (a) A number, variable, product, or quotient in a mathematical expression which does not contain a sum (+) symbol or a subtraction (-) symbol – For example, 6x is a term, but a + 2b contains two terms, or, (b) Each of the items or quantities in a sequence of items or quantities.

Tera – A metric prefix meaning a trillion or 10^{12} – For example, a trillion dollars is 1,000,000,000,000 dollars

Terminating Decimal Number – A Decimal Number that contains a finite, or countable, number of digits – A NonTerminating (or Non-Terminating) Decimal Number contains an infinite; that is, unending; number of digits. A Terminating Decimal Number is a Rational Number.

Theorem – A proposition that is to be proved or can be proved

Therefore – Logical conclusion to a logical statement or argument

Therefore symbol → ∴

Transformation – An operation which changes a mathematical expression or equation into another form

Translation – Movement of a geometric figure without rotation, such as, by sliding

Transversal of lines – A line that intersects two or more lines

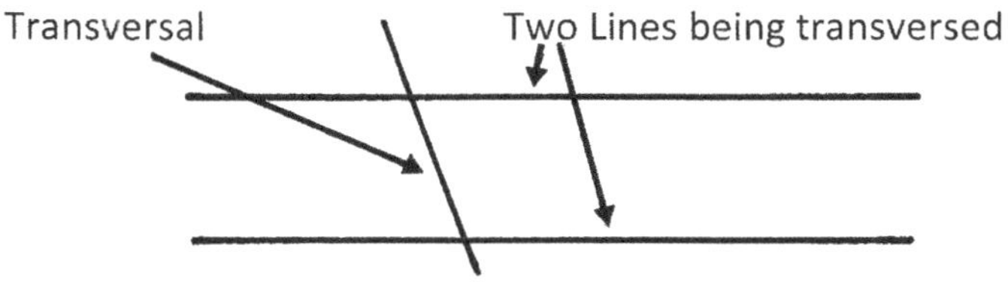

Trapezoid – a quadrilateral with only one pair of parallel sides

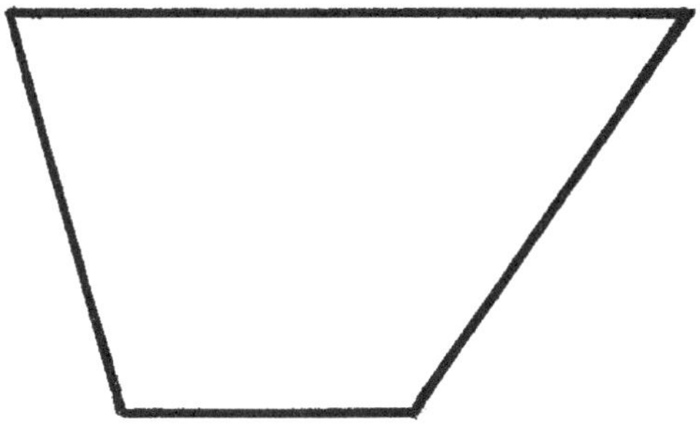

Transivity Law – For any two real numbers, a and b; if a < b and b < c; then a < c

Transitive Property – If a = b and b = c, then a = c

Triangle – A geometric figure or polygon having three angles and three sides

Trichotomy Law – For any two real numbers, a and b, exactly only "one" of the following three relationships is true: $a = b$; $a > b$; or $a < b$

Trigonometry – (a) The study of triangles and relationships between their sides and angles, or, (b) The branch of mathematics that deals with the ratios between sides of a right triangle

Truncate – (a) To make numbers easier to use by ignoring, maybe rounding off, all digits to the right of a chosen ending digit, or, (b) to delete or cut off part of a number or geometric figure or shape

Truth Table – A table showing all possible combinations of variables that are combinable when considering available combinations. A truth table also may be organized so as to show all possible outputs that may occur with all possible combinations of inputs

For example: The truth table for combining A and B wherein the combination is true only if A and B are both 1

	A = 0	A = 1
B = 0	false	false
B = 1	false	true

Undirected line – A line which does not have a direction defined for it, such as not having a direction arrow on it

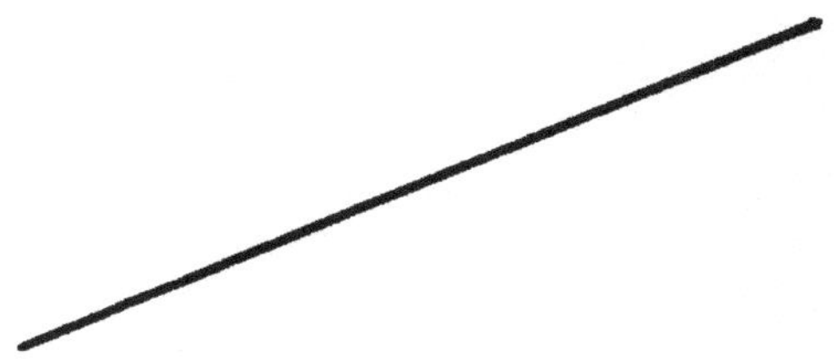

Union – A set containing all of the elements of two or more sets and no other elements

For example: Set A = 1, 3, 7, 8 and Set B = 3, 5, 9 Therefore: The Union of Set A and Set B = 1, 3, 5, 7, 8, 9

Union of two sets is frequently symbolized as A U B

Unit – (a) An item with a numerical value or magnitude of one, or, (b) a single item within a collection of items, or, (c) a standard amount or item that may be used as a comparison in counting how many items are in a collection

Unit Fraction – Any fraction with a numerator of one, "1"

Unit Vector – A special vector that has a unitary length or magnitude of one (1) in the Space in which it is defined and a Direction that may be anything within the Vector Space, however, frequently a Unit Vector has a Direction that is the same as one of the axes of the coordinate system describing the Vector Space in which the Unit Vector resides.

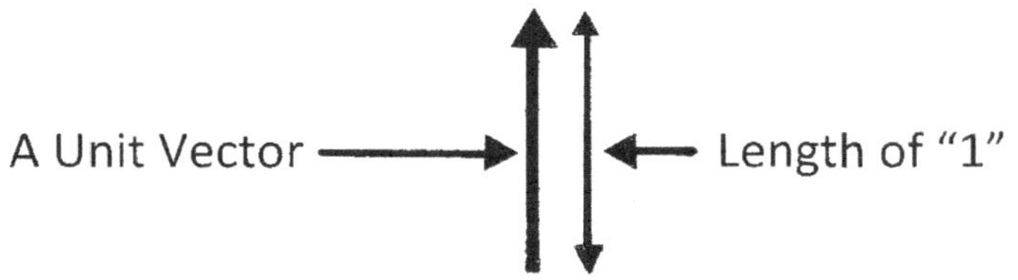

Units –A descriptive word or words for a single item within a collection of one or more items. For example, in a dozen eggs there are twelve (12) eggs. Each egg is a unit and a dozen is a collection of twelve such units or eggs. Units are words that are usually attached or written after a magnitude or number of such units. Another example is a distance of 100 miles. Miles is the units wherein each mile is the unitary single item and 100 miles indicates 100 of the units of miles. Mathematicians frequently manipulate numbers in formulas, equations, algebra, calculus, while dropping the units that should go with each number that a mathematician is manipulating in order to emphasize and simplify the mathematical manipulation, but to bring such mathematics to life and general usefulness, the units need to be reattached at some time.

Universal Set – A set from which all elements in sets being considered are taken in a specific situation

Unlike terms – Terms (plural) that do not contain the same variable or variables

Variable – Anything changeable or having no fixed value, or, a part of a mathematical expression that may assume any value in a specified, related set of values, or that may result in a value as a solution to a problem – For example, "x" and "y" are common variable symbols.

Variable Expression – (a) A mathematical expression that can have a plurality of values, or, (b) a mathematical expression that contains one or more variables – For example y = 2x

Vector – A Vector is usually represented by a line segment having both a 1) a length or magnitude or value and 2) a direction shown by an arrow on one end of the line segment

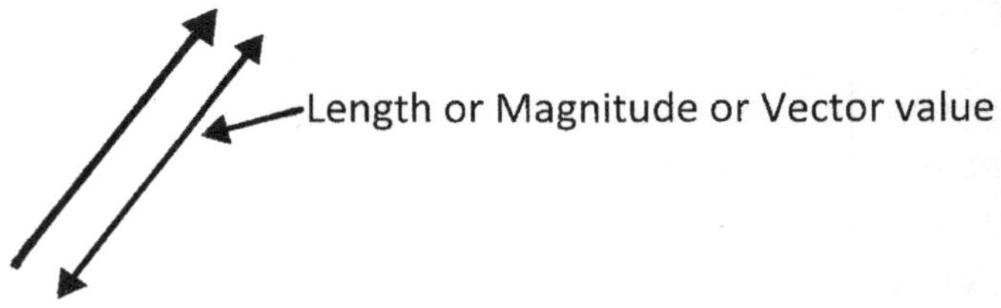

Vector Space – A Space wherein one or more Vectors can be defined. Such a Vector Space must have available therein both magnitude(s) and Direction(s) in which one or more Vectors can be defined.

Venn Diagram – A diagram using circles or other figures or shapes that may be overlapping sometimes with shading, crosshatching, or other coloration to show relationships between sets, each circle, figure, or shape representing a specific set

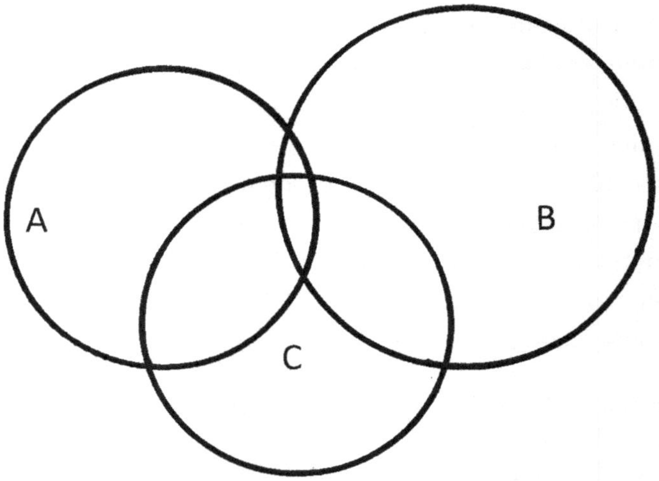

Vertex – The point at which two lines, line segments, or rays meet to form an angle

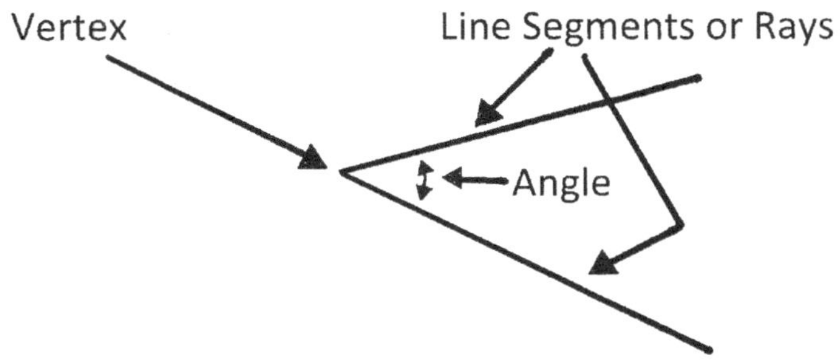

Vertices - Plural of vertex

Vertical – (a) The direction that is perpendicular to that which is horizontal, or, (b) not horizontal or level, or, (c) upright from a level plane

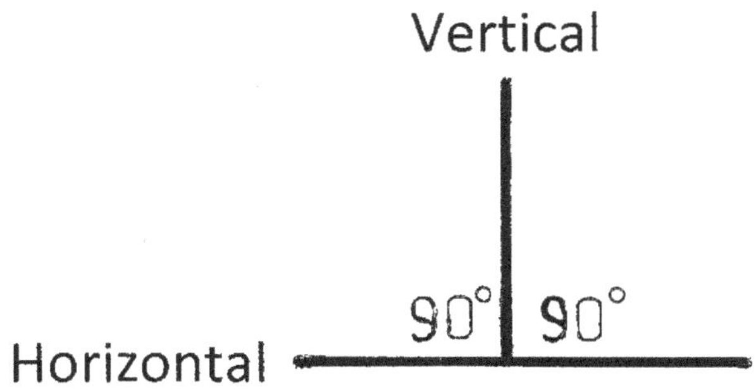

Vertical Angles – The two opposite angles that are formed when two lines cross or intersect – Vertical angles are congruent, or the same, in degrees or measurement

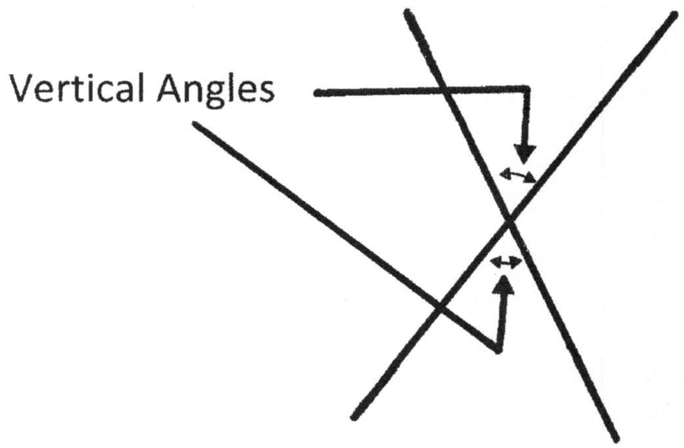

Volume – The measurement of the interior of a solid figure or shape – The volume is represented or measured in cubic number values. For example, the volume of a cube with sides being 2 feet is calculated as 2 feet times 2 feet times 2 feet which equals 8 cubic feet "8 ft^3"

Whole Number – Any of the positive integers which do not have a fractional value, as well as zero (0) – Whole numbers are 0, 1, 2, 3, 4…

X-Y plane – A plane that contains the X-axis and the Y-axis of a Rectangular or Cartesian Coordinate System

X-axis – The defining line of the X positions of points located via a Rectangular or Cartesian Coordinate System

Y-axis – The defining line of the Y positions of points located via a Rectangular or Cartesian Coordinate System

Z-axis – The defining line of the Z positions of points located via a Rectangular or Cartesian Coordinate System

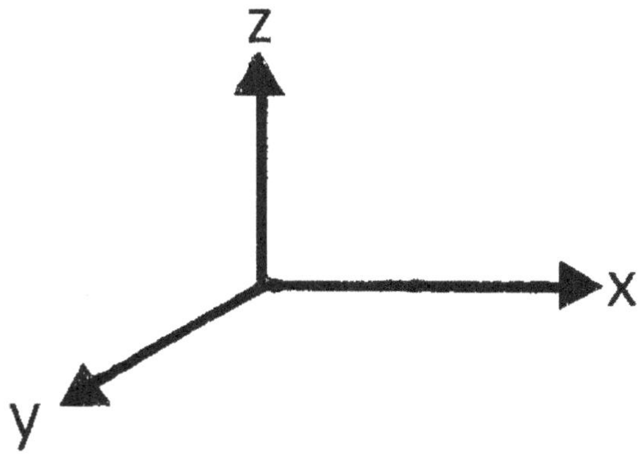

Zenith – (a) The highest point, or, (b) peak, or, (c) the point directly overhead in the sky

Zero – Zero is symbolized by "0" and indicates a null count. Zero plus the set of Natural Numbers equals the set of Whole Numbers.

Zero Property – The property or characteristic of the number zero, "0", that the multiplicative product of zero, "0", times any real number equals zero, "0": If the real number, N, is multiplied by 0; the resultant product is 0. For example: $N \times 0 = 0$ or $0 \times N = 0$ or $N \bullet 0 = 0$ or $0 \bullet N = 0$

APPENDIX A: A Very Basic Introduction to Variables and Mathematical Expressions

Two main types of variables are called dependent and independent.

Dependent variable: A dependent variable remains unknown during the start and middle of a problem. The end or solution of the problem usually is produced when the value of one or more dependent variable(s) values is(are) found.

Independent variable: An independent variable sometimes is exactly defined by the conditions in the beginning (initial) conditions of the problem. These initial conditions can vary from one problem to the next and with them the value(s) of independent variable(s). Usually, but not always, the value of an independent variable may have a known value by the conditions that are described or stated at the start of a particular problem.

Some common variable symbols are: x, y, z, u, v, w, X, Y, Z, U, V, and W

A constant is a number that is known and stays the same in value throughout the problem solving process. Some common constant are numbers, or can be symbolized such as: a, b, c, A, B, and C. Some constants are well known and used often, such as: " π " or " e ".

Besides the above variable types, solving a Math of Arithmetic problem most often starts with writing an expression, frequently as an equation, which contains variables, constants, and operation symbols. Some operation symbols are: "+" (add); "-" (subtract); "x" or "•" (multiply); "÷" or "/" (divide); " ^{2} " or " ^{3} " exponents; ( ) [ ] { }(parentheses, brackets, or braces to group subparts of expressions).

Mathematic and Arithmetic problem solving usually starts without knowing the answer until the problem is solved. The answer is usually a specific number or solution to what is desired to find. Such solution answers therefore usually start as unknowns, usually represented by symbols, called variables. Sometimes in special situations, a variable can be known as starting information but may be different for different problems.

An example of a Mathematical Expression is: Y = 2 • X + B. A simple Math problem would be to find the value (or solution of the problem) of "Y". "Y" is a dependent variable. "•" and "+" are math operation symbols. "X" is an independent variable. "B" is a constant. If X equals 2 and B equals 3, then the next step is to replace (or substitute values for) X and B in the Mathematical Expression. This results in "Y" being equal to "2 • 2 + 3", which simplifies to "4 + 3". Note that there is an order to doing math operations in simplifying an expression. Multiplying, symbolized here by " • ", is done before addition, symbolized here by " + ". Further simplification then results in finding that Y equals 7. This value of 7 for Y is the solution to the equation.

APPENDIX B: A Very Basic Introduction to Solving Mathematical Problems

A. Solutions or Answers to most Mathematics problems are exact values or numbers.

Most teachers and tests want solutions or answers in their simplest form. Money problems usually have an answer in dollars and cents. Geometry problems may have an answer in a length, area, or volume of a geometric figure. Some Mathematics problems are very similar to each other and a general or common answer can be figured out as a formula or expression for such similar problems.

B. An example problem and solution using dependent and independent variable(s)

A simple problem using variables might be to sum up the number of fruits, such as apples and oranges, purchased while going to a grocery store. Different numbers of apples and oranges might be purchased at different visits to the grocery store, so solving this problem may be useful again and again. So let's define the number of Apples as the initial independent variable value: "A". The initial number of Oranges is symbolized by the independent variable: "O". The problem is to find out how many total Fruits are purchased in a particular grocery visit. So let's mathematically symbolize this final desired answer by the dependent variable: "F". The variable "F" is a dependent variable because its value depends on the other variables, "A" and "O".

Therefore, the answer regarding the number of total purchased fruits is found or calculated by adding the number of Apples to the number of Oranges purchased. Therefore, the solution to this problem is symbolized mathematically by the following mathematical sentence which is also known as an expression:

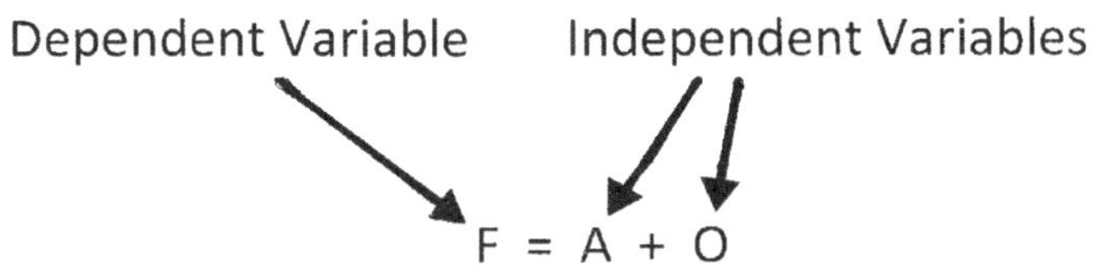

This expression is shorthand for writing that "the total number of Fruits purchased equals the number of Apples purchased added to the number of Oranges purchased". $F = A + O$ is a lot easier to write than all those words.

Let's solve for the total number of fruits purchased in a particular grocery store visit. In this visit, 3 Apples and 2 Oranges are bought. Therefore, the independent variable A has the value of 3. The independent variable O is 2. So we set A to a value of 3 and O to a value of 2 in our Math expression and calculate the dependent F variable result by using the formula or expression: $F = A + O$. For this visit,

F = 3 + 2 and the answer is 5. 5 Fruits were purchased. So now we know at the end of this problem that the dependent variable "F" answer is 5 Fruits. The dependent variable value depended on the values of the two independent variables, A and O.

Please notice that the numbers; 3, 2, and 5 themselves are abstract and meaningless without knowing that each actually is connected to something real such as the number of Apples, Oranges, and Fruits, respectively. These connected real ideas are called "units" in Mathematics. These "units" make abstract numbers mean something in real life. We assumed during the above grocery visit problem that we kept in mind that A actually carried the meaning with it of the number of Apples and O carried with it the number of Oranges. During the calculation, however, we temporarily dropped writing those units with the number values. We assumed, or hoped, that we would remember the unwritten units connected with each variable.

StudyKat

"Meow" Lots of Math books contain Math problems and exercises using only numbers, symbols, and variables without units like Apples or Oranges. Please remember that the Math shorthand helps to solve problems easier with practice but that all of them in real life usually have real life units and uses.

If a more complicated Mathematical problem was being worked on and solved, should we continue writing the units during the problem solving to avoid confusion as to what each variable means? Maybe. Maybe not. The handling of written units, versus dropping them, during Mathematical problem solving is tricky, and a personal preference. Keeping the units with numbers takes work and is usually takes a lot of time, so dropping them in the middle of calculations is a time saver, "as long as you can remember what is meant". Mathematicians frequently wish to focus problem solving on the Pure Mathematics of numbers and Math operations only and therefore drop units for efficiency and simplicity. This can make Mathematics somewhat unreal looking and unconnected to useful real life problems. However, many areas of Mathematical problem solving methods were born from real life problems that needed solving. At the time, previous Mathematical problems solving processes were just not good enough.

C. Lastly, a few more famous Constants

Any quantity or item that has the same value over many different situations is known as a constant. Some constants get used so often that their symbols become very well known.

Some well known constants are as follows:

The number π (pronounced "pie"), has the value: 3.1415926536 . . . (sometimes rounded to 3.14)

The speed of light is symbolized by “c” and has the value: about 186,000 miles per second

The natural logarithm base, symbolized by “e”, has the value: 2.71828 . . . (sometimes rounded to 2.72)

Many other constants are known and are useful. Some of the more common ones are listed and briefly explained in the above Definitions/Glossary.

<u>APPENDIX C:</u> A Very Basic Introduction to Using Fractions and Fractional Numbers

Before getting to Fractions or Fractional Numbers, Whole Numbers and Integers need to be understood first. The above Part I listed some Whole Numbers already. Integers are whole numbers that include zero and negative numbers, as well. A Natural number is any positive Integer.

Natural Numbers: 1, 2, 3, 4, 5, 6, 7, 8, 9, 10, 11, 12, 13, 14, . . .

Whole Numbers: 0, 1, 2, 3, 4, 5, 6, 7, 8, 9, 10, 11, 12, 13, 14, . . . [Whole Numbers equals Natural Numbers plus zero (0)]

If we add a negative number for each Natural Number,+ and add zero (0), we get Integers.

Integers: . . . -12, -11, -10, -9, -8, -7, -6, -5, -4, -3, -2, -1, 0, 1, 2, 3, 4, 5, 6, 7, 8, 9, 10, 11, 12, . . .

Here is a drawing of a "Number Line" showing Integers:

. . . -3 -2 -1 0 1 2 3 . . .

Negative Infinity ⟵⟶ Positive Infinity

Located between these Integers are fractions or fractional numbers. Fractions or Fractional Numbers are introduced here with some examples.

Examples of fractions or fractional numbers:

11/13 is a fraction

Written in a longer way, the above fraction is: $\dfrac{11}{13}$

Written in an even longer way the above fraction is: "Eleven divided by Thirteen", or, Eleven over Thirteen".

88/93 is a fraction

-5/6 is also a fraction

456/12 is fraction (This is also called an Improper fraction. It is an improper fraction because the top number, 456, is greater than bottom number, 12.

In the above fractions, the top number is called "numerator" and the bottom number is called "denominator".

Types of fractional numbers, defined in the above Definitions/ Glossary, may be Common, Simple, Proper, Improper, Complex, Compound, Undefined, Mixed, Unit Fractions, or Decimal Fractions.

Decimal fractional number examples are important because they are used a lot:

0.3749 is a decimal fraction (a number containing a decimal point to separate the Integer number, left-hand part: "0", from the fractional, right-hand, part: ".3749"). The fractional part is always valued at less than the Whole number "1".

12.89605 is a number that contains a decimal fraction [(a number containing a decimal point to separate the Integer number (left-hand part: "12") from the fractional (right-hand, part: ".89605")].

Mathematical handling of fractions and fractional numbers will not be described further in this book. Many situations involve detailed explanations to handle them correctly. It therefore deserves its own informational book.

An added note: Fractions are also useful because a type of number called a Rational Number is a number that can be written as a fraction (or fractional ratio) with integers in both the numerator and denominator. For example, the number 5 can be written as 5/1 and therefore "5" is a not only a Natural Number, Whole Number, and Integer, but is also a Rational number. If a number cannot be written, transformed, or simplified into a fraction of Integers both in the numerator and denominator, the number is called an Irrational number. For example, the square root of "2" equals 1.41421 . . . (a non-terminating and non-repetitive number) and cannot be written as a fraction of Integers and is an Irrational number. A last interesting note is that a decimal fraction that is completely made up of repeats of one or more numbers in a row can be transformed into a ratio of integers and is a Rational Number.

END